T0328331

BEST:

IMPLEMENTING CAREER DEVELOPMENT ACTIVITIES FOR BIOMEDICAL RESEARCH TRAINEES

BEST:
IMPLEMENTING CAREER DEVELOPMENT ACTIVITIES FOR BIOMEDICAL RESEARCH TRAINEES

Edited by

LORENA INFANTE LARA
Vanderbilt University, Nashville, TN, United States

LAURA DANIEL
Vanderbilt University Medical Center, Nashville, TN, United States

ROGER CHALKLEY
Vanderbilt University, Nashville, TN, United States

ACADEMIC PRESS
An imprint of Elsevier

ELSEVIER

Academic Press is an imprint of Elsevier
125 London Wall, London EC2Y 5AS, United Kingdom
525 B Street, Suite 1650, San Diego, CA 92101, United States
50 Hampshire Street, 5th Floor, Cambridge, MA 02139, United States
The Boulevard, Langford Lane, Kidlington, Oxford OX5 1GB, United Kingdom

Notices
Knowledge and best practice in this field are constantly changing. As new research and experience broaden
our understanding, changes in research methods, professional practices, or medical treatment may become
necessary.

Practitioners and researchers must always rely on their own experience and knowledge in evaluating and
using any information, methods, compounds, or experiments described herein. In using such information or
methods they should be mindful of their own safety and the safety of others, including parties for whom
they have a professional responsibility.

To the fullest extent of the law, neither the Publisher nor the authors, contributors, or editors, assume any
liability for any injury and/or damage to persons or property as a matter of products liability, negligence or
otherwise, or from any use or operation of any methods, products, instructions, or ideas contained in the
material herein.

Library of Congress Cataloging-in-Publication Data
A catalog record for this book is available from the Library of Congress

British Library Cataloguing-in-Publication Data
A catalogue record for this book is available from the British Library

ISBN: 978-0-12-820759-8

For information on all Academic Press publications visit our website at
https://www.elsevier.com/books-and-journals

Publisher: Mica Haley
Acquisitions Editor: Stacy Masucci
Editorial Project Manager: Kristi Anderson
Production Project Manager: Swapna Srinivasan
Cover designer: Christian J. Bilbow

Typeset by TNQ Technologies

Contents

14. Implementation of a career cohort model at UNC Chapel Hill: Benefits to students, programs, and institutions

Rebekah L. Layton, Patrick D. Brandt and Patrick J. Brennwald

15. Vanderbilt's ASPIRE program: Building on a strong career development foundation to change the Ph.D.-training culture

Kimberly A. Petrie, Ashley E. Brady, Kate F.Z. Stuart, Abigail M. Brown and Kathleen L. Gould

16. VT-BEST: Shaping biomedical professional development programming across colleges and campuses

Audra Van Wart and Michael J. Friedlander

17. Across disciplines: Multi-phase career preparation for doctoral students

Christine S. Chow, Ambika Mathur and Judith A. Moldenhauer

Contributors

Janet Alder School of Graduate Studies, Rutgers University, Piscataway, NJ, United States

Avery August Cornell BEST Program-Careers Beyond Academia, Cornell University, Ithaca, NY, United States

Tracey Baas University of Rochester, Rochester, NY, United States

Chelsea R. Barbercheck BU's BEST, Department of Medical Sciences & Education, Boston University School of Medicine, Boston, MA, United States; Great Plains IDEA, Kansas State University, Manhattan, KS, United States

Lars Berglund Clinical and Translational Science Center, University of California, Davis, CA, United States

Amanda Florence Bolgioni BU's BEST, Department of Medical Sciences & Education, Boston University School of Medicine, Boston, MA, United States

Ashley E. Brady Biomedical Research and Education Office, Vanderbilt University, Nashville, TN, United States

Patrick D. Brandt Office of Graduate Education, University of North Carolina at Chapel Hill, Chapel Hill, NC, United States

Patrick J. Brennwald Department of Cell Biology & Physiology, University of North Carolina at Chapel Hill, Chapel Hill, NC, United States

Abigail M. Brown Biomedical Research and Education Office, Vanderbilt University, Nashville, TN, United States

Roger Chalkley Vanderbilt University, Nashville, TN, United States

Christine S. Chow Department of Chemistry, Wayne State University, Detroit, MI, United States

Rebekah St. Clair School of Public Policy, Georgia Institute of Technology, Atlanta, GA, United States

Milagros Copara Clinical and Translational Science Center, University of California, Davis, CA, United States

Tamara Dahl Laney Graduate School, Emory University, Atlanta, GA, United States

Laura Daniel Vanderbilt University, Nashville, TN, United States

Steve Dewhurst University of Rochester, Rochester, NY, United States

M. Isabel Dominguez Department of Medicine, Boston University School of Medicine, Boston, MA, United States

Jennie Dorman University of California, San Francisco, CA, United States

Susan R. Engelhardt Center for Innovative Ventures of Emerging Technologies, Department of Biomedical Engineering, Rutgers University, Piscataway, NJ, United States

Spencer L. Fenn Graduate School of Biomedical Sciences, University of Massachusetts Medical School, Worcester, MA, United States

Michael J. Friedlander Fralin Biomedical Research Institute, Roanoke, VA, United States; Virginia Tech Carilion School of Medicine, Roanoke, VA, United States; Department of Biological Sciences, Virginia Tech, Blacksburg, VA, United States

David A. Fruman Department of Molecular Biology and Biochemistry; GPS-BIOMED, University of California at Irvine, Irvine, CA, United States

Cynthia N. Fuhrmann Graduate School of Biomedical Sciences, University of Massachusetts Medical School, Worcester, MA, United States

Kathleen L. Gould Biomedical Research and Education Office, Vanderbilt University, Nashville, TN, United States

Jennifer Greenier Clinical and Translational Science Center, University of California, Davis, CA, United States

Stacy Hayashi Clinical and Translational Science Center, University of California, Davis, CA, United States

Daniel Hidalgo Graduate School of Biomedical Sciences, University of Massachusetts Medical School, Worcester, MA, United States

Sarah Chobot Hokanson Office of the Provost, Boston University, Boston, MA, United States

Brent B. Horowitz Graduate School of Biomedical Sciences, University of Massachusetts Medical School, Worcester, MA, United States

Linda E. Hyman Graduate Medical Sciences, Boston University School of Medicine, Boston, MA, United States; Marine Biological Laboratory, Woods Hole, MA, United States

Arthee Jahangir Office of Postdoctoral Affairs, New York University Grossman School of Medicine, New York, NY, United States

Karen Klomparens The Graduate School, Michigan State University, East Lansing, MI, United States

Mary Ellen Lane Graduate School of Biomedical Sciences, University of Massachusetts Medical School, Worcester, MA, United States

Lorena Infante Lara Vanderbilt University, Nashville, TN, United States

Rebekah L. Layton Office of Graduate Education, University of North Carolina at Chapel Hill, Chapel Hill, NC, United States

Bill Lindstaedt University of California, San Francisco, CA, United States

Heather S. Loring Graduate School of Biomedical Sciences, University of Massachusetts Medical School, Worcester, MA, United States

Ambika Mathur Department of Pediatrics, Wayne State University (previous), Detroit, MI, United States; University of Texas, San Antonio, TX, United States

Nael A. McCarty Laney Graduate School, Emory University, Atlanta, GA, United States; School of Medicine, Emory University, Atlanta, GA, United States

Julia Melkers School of Public Policy, Georgia Institute of Technology, Atlanta, GA, United States

Frederick Meyers Clinical and Translational Science Center, University of California, Davis, CA, United States

Keith Micoli Office of Postdoctoral Affairs, New York University Grossman School of Medicine, New York, NY, United States

Daniel Moglen Clinical and Translational Science Center, University of California, Davis, CA, United States

Judith A. Moldenhauer Department of Art and Art History, Wayne State University, Detroit, MI, United States

Gabriela C. Monsalve University of California, San Francisco, CA, United States

Sumeet Nayak Graduate School of Biomedical Sciences, University of Massachusetts Medical School, Worcester, MA, United States

Wendy C. Newstetter College of Engineering, Georgia Institute of Technology, Atlanta, GA, United States

Theresa C. O'Brien University of California, San Francisco, CA, United States

Kimberly A. Petrie Biomedical Research and Education Office, Vanderbilt University, Nashville, TN, United States

Sarah Peyre University of Rochester, Rochester, NY, United States

Christine Ponder Research Affairs, Postdoctoral Affairs, New York University, New York, NY, United States

Rachel L. Reeves Clinical and Translational Science Center, University of California, Davis, CA, United States

Carol Shoshkes Reiss Department of Biology, New York University, New York, NY, United States

Julie W. Rojewski The Graduate School, Michigan State University, East Lansing, MI, United States

Chris B. Schaffer Cornell BEST Program-Careers Beyond Academia, Cornell University, Ithaca, NY, United States

Barbara M. Schreiber Department of Biochemistry, Boston University School of Medicine, Boston, MA, United States

Elizabeth A. Silva University of California, San Francisco, CA, United States

Harinder Singh GPS-BIOMED, University of California at Irvine, Irvine, CA, United States

Meghan E. Spears Graduate School of Biomedical Sciences, University of Massachusetts Medical School, Worcester, MA, United States

Jean L. Spencer Department of Biochemistry, Boston University School of Medicine, Boston, MA, United States

C. Abigail Stayart Biological Sciences Division, University of Chicago, Chicago, IL, United States

Kate F.Z. Stuart Biomedical Research and Education Office, Vanderbilt University, Nashville, TN, United States

Audra Van Wart Fralin Biomedical Research Institute, Roanoke, VA, United States; Virginia Tech Carilion School of Medicine, Roanoke, VA, United States; Division of Biology and Medicine, Brown University, Providence, RI, United States

Susi Varvayanis Cornell BEST Program-Careers Beyond Academia, Cornell University, Ithaca, NY, United States

Bineti Vitta Clinical and Translational Science Center, University of California, Davis, CA, United States

Stephanie W. Watts The Graduate School, Michigan State University, East Lansing, MI, United States

Grant C. Weaver Graduate School of Biomedical Sciences, University of Massachusetts Medical School, Worcester, MA, United States

Inge Wefes University of Colorado Denver, Anschutz Medical Campus, Aurora, CO, United States

Keith R. Yamamoto University of California, San Francisco, CA, United States

Origin of BEST, and how 17 very different programs created 17 related approaches to help trainees with their career choices

Laura Daniel, Lorena Infante Lara, Roger Chalkley
Vanderbilt University, Nashville, TN, United States

What is BEST?

The Broadening Experiences in Scientific Training (BEST) program was established with support from the National Institutes of Health (NIH). The goal of the program has been to test strategies to support and prepare biomedical research graduate students and Ph.D.'s in a wide range of careers. The support consisted of non-renewable five-year grants that were awarded in 2013 (to ten institutions) and in 2014 (to seven institutions).

This book will focus on the experience of the BEST institutions; however, it is important to recognize that a number of schools who are not a part of the BEST Consortium have made similar career development efforts.

A call for graduate education reform in biomedical sciences

Traditionally, apart from the efforts of individual faculty, there had been no organized system to instruct graduate students and postdocs in the biomedical sciences about the range of career options available to them. A trainee was expected to go into academia, with industry a more or less acceptable alternative. Otherwise, they were on their own.

One of the earliest attempts to think about trainee career needs came from April Hamel, Associate Dean of the Graduate School at Washington University in St. Louis, who conducted a survey of the career services other universities were providing for their Ph.D. students in 1987. This marked the beginning of the Graduate Career Consortium (GCC), an organization whose mission is to support career and professional development for doctoral students and postdoctoral scholars, and which has grown in importance, influence, and scope and includes over 400 members as of 2018 [1].

A parallel attempt to refocus graduate and medical education occurred in the mid-1990s with the formation of another group concerned with career training and preparation, the Graduate Research, Education, and Training (GREAT) Group. The GREAT Group, sponsored by the American Association of Medical Colleges, is an organization for faculty and administrative leaders dedicated to graduate and postdoc education that meets annually. At an early meeting, a group of postdocs and graduate students relayed their issues and frustrations to a group that seemed quite unaware that postdocs had any problems at all.

This triggered a series of changes, driven to a considerable degree by the students and postdocs themselves, including the conceptualization of a national organization envisioned and designed specifically to serve the needs of postdocs. The recognition of the issues common to postdoctoral scholars also influenced several institutions that realized that it was time to begin to support the notion of training for careers. In 1995, *Science* published a special issue discussing the "crisis" of a growing population of Ph.D. scientists and a stagnant pool of faculty jobs. A debate ensued about whether to curtail the production of Ph.D.'s or to make graduates "more marketable in an uncertain future" [2]. *Science* examined the concern from the point of view of all stakeholders, which marked the beginning of the efforts of the journal and AAAS in addressing the career needs of young scientists by offering a new resource, *NextWave*; this weekly web magazine eventually became *Science Careers*.

Although there is little documentation about the individual steps that were taken immediately following the GREAT Group meeting, one of the first career-centric symposia within the biomedical sciences was held at Vanderbilt University, a future NIH BEST award recipient. This inaugural symposium took place over a day and a half in the summer of 1997. The first day was focused on what are called "traditional" careers today, although it introduced some panelists and speakers who focused on teaching rather than research and others who hailed from biotech-like companies; the presence of this latter group was something rather novel, as industry representatives typically had not attended this type of event in the past. The second day was dedicated to opportunities for visiting professionals to meet with students or postdocs one-on-one to discuss what was a much broader career landscape than the one trainees were most familiar with. The event was such a success that the university began to host the Career Symposium on a regular basis, with the following event set for three years hence. The extended timeline adopted is perhaps a stark indication of the lack of urgency on behalf of training programs at that time.

The next ten years were, in a sense, a period of incubation for the changes that would begin to solidify in the early 2010s. GCC continued to be a vibrant community across this time period, with an active listserv and an annual meeting through which ideas were exchanged amongst professionals serving the career development needs of Ph.D. students and postdocs. The National Postdoctoral Association (NPA) [3] was established in 2003 following a *NextWave* postdoc network meeting. Postdoc offices, whose sole focus was to manage postdoc affairs, began to appear across the nation, especially in the biomedical research arena. In fact, as discussed in a few of the chapters, some BEST programs were built upon career information and programs that came out of a local postdoc office. During this time period, the GREAT Group introduced regular discussions of different careers at its annual meetings, and asked postdoc attendees what professional development support they would like institutions to provide. The discussions again revealed the mismatch between needs and programming, highlighting how little many institutions were doing for their trainees in this arena.

It was evident that there were not enough new tenure-track faculty positions to assure postdocs who were completing their training that they could have a decent chance of landing a faculty job. Although this was nothing new, in the absence of good career development support, 43% of postdocs felt they were not receiving proper training and 24% stated that they didn't view their advisor as a mentor [4]. In addition, there was a disconnect between the percent of postdocs who expected to obtain a tenure-track faculty position (56%) and those that actually did (30%) [5]. Postdocs, especially those at elite research institutions, felt that if they had dutifully invested their time then they should be assured of a prestigious faculty position. Of course, the situation was never thus, but sometimes sensationalized reporting in scientific magazines amplified the frustrations of postdocs who felt that in signing on as graduate students they had been guaranteed of a career just like that of their advisors [6,7]. This led to reports of an exponential increase in trainees with no jobs available for them. It was not until 2016–17 that careful analyses of the National Science Foundation (NSF) databases revealed that this trend had tapered off such that after 2008 there were no longer increases in the number of Ph.D.'s who graduated from biomedical science programs [8,9].

Several schools introduced their own career symposia, career panels, workshops, or other programs during the decade starting in 2000. Graduate student and postdoctoral associations established career-focused clubs and other activities on their local campuses. The content progressively broadened to include an ever-wider range of careers,

especially as the number of graduates of the increasingly popular umbrella programs (centralized graduate programs that span multiple departments and/or programs) rose rapidly. By the mid-2000s, many schools were moving towards offering career symposia on an annual basis in response to strong interest from graduate students and postdocs alike.

Around this time, it was clear to anyone involved with graduate/postdoc training at competitive institutions that in order to maintain leadership in research education, they had to move aggressively into new approaches for biomedical research training. Several institutions (including some soon-to-be BEST institutions) invested in career development and in recording the outcomes of their trainees during this time period. A few of these schools could provide a reasonable record of the overall types of careers their graduates entertained by the end of the decade, but this was far from the norm at the national level. Thus, as the decade drew to a close, it became clear that most biomedical science training approaches were not sufficiently addressing the career development needs and interests of Ph.D. trainees. The early 2010s saw nationwide recognition that something more had to be done to improve and update the education process that had changed little in the previous 50 years.

In 2011, the NIH called together a blue-ribbon committee, the Biomedical Research Workforce Working Group [10]; to suggest strategies to deal with the perceived over-production of biomedical research Ph.D.'s. After a series of meetings and hearings with stakeholders, this group proposed a rather thoughtful and modest course of action. One of the main recommendations was to improve career training for graduate student and postdoc trainees in the biomedical sciences with an eye towards broadening the career landscape for such newly minted doctoral graduates [11].

The NIH responded promptly to the recommendations from this committee and released a funding opportunity announcement (FOA), NIH Director's Biomedical Research Workforce Innovation Award: Broadening Experiences in Scientific Training (BEST) (DP7), that stated:

> *The purpose of this FOA is to seek, identify and support bold and innovative approaches to broaden graduate and postdoctoral training, such that training programs reflect the range of career options that trainees (regardless of funding source) ultimately may pursue and that are required for a robust biomedical, behavioral, social and clinical research enterprise. Collaborations with non-academic partners are encouraged to ensure that experts from a broad spectrum of research and research-related careers contribute to coursework, rotations, internships or other forms of exposure. This program will establish a new paradigm for graduate and postdoctoral training; awardee institutions will work together to define needs and share best practices [12].*

Each institution was to take its own approach, experimenting with ways to best provide support for their trainees. The NIH expected that some of the approaches would work well, whereas others would not; collectively, these experiments were to provide guidance for other institutions in the future. As a research grant, the NIH required each site to have its own evaluation strategy and to contribute to a cross-site evaluation that would be centrally coordinated. A total of 17 awards were made, each consisting of a one-time grant with no prospect of renewal—a feature that encouraged the awardees to hit the ground running with a goal of creating programs that could be sustained outside the long-term grant mechanism. A concern for these institutions was whether or not the faculty would be receptive and supportive of these kinds of activities. Thus, it became necessary to address the need for cultural change at each institution, and this became a fundamental part of the challenges they faced in developing and implementing their programs.

Getting started

The 17 schools that received BEST awards were amazingly different. They ran the gamut from small to large schools, from private to public institutions, and from medical schools to comprehensive universities. In many cases, the biomedical research programs accounted for the major source of students and postdocs involved in research at a particular institution; in others, research in the biomedical sciences contributed only a relatively small fraction of the student numbers, which was often the case in larger institutions with substantial STEM research; for instance, Michigan State University has 4,700 graduate students, but only ~70 in biochemistry and molecular biology.

The awards were announced in the late summer of 2013 and 2014 and were awarded rapidly after review, leaving the recipients with relatively little forewarning of their good fortunes. This tended to favor those programs that had already started to develop approaches for broadening their career training offerings, as they were able to expand rapidly on ongoing successes and capitalize on faculty support to allow for inventive new strategies.

There were two main implementation methods: institutions either hired a program director to manage the day-to-day activities of their program (oftentimes a relatively recent postdoc already present at the institution), or they used some of the funding to buy-out faculty time so that one or two faculty members could spend more time developing and establishing the programs that the institutions had outlined in their applications to the NIH. Some programs pursued a combination of these approaches. Today, institutions looking to found their own career development offices or programs are likely to find personnel that already has some experience in this arena.

Many of the programs that received BEST funding organize their training of graduate students through umbrella programs. In these cases, it was advantageous for the BEST-funded career development staff to be located close to the central office that supports the graduate program, as it allowed convenient access to the students and an opportunity to interdigitate career and graduate training seamlessly to the maximum degree possible. This is not to say that separation from the student hubs would make the programs less functional, but it would likely impose extra difficulties on the staff in the career development office.

All of the funded institutions were affiliated with a medical school except for Cornell University. Given the traditional organizational structure of graduate programs in medical schools, the association with medical schools was probably beneficial for two reasons. First, such graduate programs have a long history of a reasonable degree of independence. Second, they have been traditionally well supported by NIH training grants, which stress innovative programming and emphasize career development. As a result, medical schools have been inclined to help underwrite their basic biomedical science Ph.D. career development activities.

Comprehensive universities (either with or without medical schools) also bring advantages, such as having other programs/resources available (writing centers, business schools, etc.), which can benefit the establishment of numerous BEST-type enrichment activities.

From the beginning, the BEST institutions have strived to follow the principle that instruction and career guidance should be equally accessible to graduate students and postdocs. In addition, the various programs have devised mechanisms to generate different career discussions at appropriate stages in the trainees' careers.

Dispelling the myth of the unsupportive faculty

On discussing the various start-up approaches of all the BEST programs, one thing was critical for their success: communication—especially with faculty but also with trainees. In order to establish broad understanding of the new programs, most leadership teams contacted faculty promptly following the receipt of the award to explain what BEST meant and what they hoped to achieve. Some schools started with a campus-wide launch event and email and word-of-mouth communication campaigns.

A strategy that seems to have worked well was for a program leader to attend individual departmental faculty meetings. The potent logic behind this approach was the widespread perception that the faculty would be opposed to anything that took trainees out of the lab even for an occasional hour or 2 per week, and certainly for the 4 to 12 weeks entailed in pursuing internships (an experience many BEST groups hoped to offer). Faculty hesitance was a deep concern discussed among all the BEST programs in the early days. Encouragingly, many of the BEST programs reported the identification of "faculty champions"—chairs, seasoned administrators, or senior faculty—who recognized the need for implementing changes in how trainees were educated; these supporters were helpful in arguing the case of BEST from a peer-to-peer perspective with the few potential naysayers.

The notion of the profoundly disapproving faculty arose from reports in national journals in which individual postdocs had been asked for their opinions in ways that did not ensure that those opinions were representative of the population as a whole. Of course, the false perception of the exploited trainee made popular press, but it clearly bore little resemblance to the reality on the ground. Only a small portion of postdocs felt that their advisor was unsupportive [13].

Two early surveys on faculty responses to BEST were combined and have been recently published, indicating much more favorable responses than previously popularized [14]. Findings indicated that most faculty were well aware of trainees' job situations and that they realized that the likelihood of an individual trainee ending up in a tenure-track faculty position at an R1 (research) institution was small. However, the survey also determined that faculty were largely unaware of other career possibilities available to well-trained Ph.D. scientists outside the academy, and that they did not feel prepared to mentor trainees for these unknown careers. The BEST programs found that the faculty readily admitted their lack of knowledge and that in most cases they were pleased that their institutions provided programs that helped support their students and postdocs in becoming better informed in these other areas. Additional enthusiasm from the faculty derived from the BEST programs' ability and willingness to prepare trainees for such careers while balancing expectations for trainees' time in the lab. The surveys indicated that most faculty felt that two to four hours per month was not an unreasonable investment for the student to make in their career development. Additionally, informal internal surveys at several BEST institutions indicated that early efforts in the award period were recognized and even appreciated by many of the faculty.

By the midway point in the BEST programming, it was clear that faculty were, for the most part, quite supportive of the BEST initiatives. Graduate education is essentially the ultimate *quid pro quo*, wherein the agreement is that if the student works and focuses effectively, then the faculty will create the guidance and funding for successful research training. But this arrangement is only complete if the training environment is able to ensure that the trainee has a fair chance at developing a career that is satisfying and rewarding. The BEST program has been a major driver in creating opportunities to improve the training experience for all its participants.

One thing that BEST program administrators discovered early on was that faculty sometimes feared that the students and postdocs were being specifically trained and encouraged towards non-faculty positions as evidenced by the emphasis and training on the wide range of career opportunities. Although this was never the goal of BEST, all the programs learned to make especially transparent efforts to assure faculty that all career pathways were being supported, including academic and research-intensive careers like their own.

Fundamental aspects of the structure and content of the BEST programs

Trainees have been some of our major supporters, a sign that the programming is perceived as useful and relevant.

Unpublished surveys, including the NIH cross-site survey (see below), have provided evidence that this is the case. To a major degree, this reflects the content of the programs the various BEST groups have developed. Although all the institutions have developed their own idiosyncratic content, many commonalities have emerged. As a result, trends and popular program options developed organically in similar ways to cover the material needed to support the trainees. Thus, although there were many different approaches, the programs now share much in common despite very specifically tailoring their offerings to their own needs and trainee interests.

Trainees across BEST programs are introduced to a wide range of career options and to ways of exploring them. At an early stage, most programs also introduce trainees to the concept, goals, and use of an Individual Development Plan (IDP). By the end of the first year in graduate school, programs hope that BEST trainees are at least exposed to a range of career options, and that trainees are aware of how their program can help them as they begin to plot their future careers. As students progress through graduate training, they are encouraged to invest more extensively in career exploration and preparation as they mature into senior graduate students and eventually Ph.D. candidates (graduate students do not become Ph.D. candidates until they pass a qualifying exam). Although these are common themes, the approaches that institutions have taken to address these needs have varied.

Recent findings also indicate that although the students participate in the process of career exploration throughout their training, their time to degree is not impacted [15] and faculty do not believe that it is a major and continuing distraction from their research activities [16].

Workshops, seminars, and courses

BEST programs commonly provide recurring seminars, career panels, or small-group informal discussions on specific career paths given by different Ph.D.-level professionals, often an alumnus or alumna, recounting their own career paths. Doctoral career outcome data collection (see below) has greatly facilitated locating relevant alumni who hold positions in the featured careers and who can give appropriate guidance to trainees. These presentations, combined with informal Q&A sessions, allow the listeners to tailor their questions to their own interests. Having regular programming like this helps establish that it is acceptable to discuss and consider a range of career options that may not be traditional in any sense. A few programs have emphasized this by building these opportunities into the required graduate and/or postdoctoral curriculum. Some programs provide support and encourage trainees to play a major role in selecting speakers and focus areas, while others primarily use staff to take the lead in contacting alumni (or other speakers) to participate.

Another popular strategy is to create workshops focused on specific skills (e.g., teaching, coding, big data analysis, science writing, etc.), which may run on a regular basis for several weeks. This approach is often coupled with a registration and recording process to follow attendance. Some schools go even further and accommodate a more structured approach, creating a course that is integrated into graduate student training and for which students may receive a grade that goes on their transcript. Additionally, many schools have identified (local) faculty with specific knowledge about certain career areas and a commitment to graduate education who can contribute course content; these faculty are often willing to provide this expertise as part of their institutional service.

Early on, several programs identified external companies or consultants who were paid to give in-depth training in specific areas. However, it became clear that this approach, while easy to put in place and valued for its expertise and professionalism, could be expensive. After seeing the content and style of such presentations it was natural for some BEST staff to use homegrown material to come up with their own variations on a theme, and their presentations have shifted in this direction over the years; often, this has evolved to include other BEST programming that can be presented elsewhere. Nonetheless, there continues to be value in inviting external speakers (not infrequently alumni) with different experiences, perspectives, and specialty areas to present. Many institutions provide a balance of internal and external speakers to vary material and topics.

Experiential learning

Externships and internships have played a role in the BEST experience since they were initially discussed at the first annual meeting in 2013. At the time the awards were made, it was probable that very few attendees knew much about these activities; in fact, many of the BEST program administrators did not understand the difference between internships and externships, let alone how to set them up. The University of North Carolina at Chapel Hill had the somewhat rare experience of creating a new internship program from the ground up in the first year of the BEST funding. This process included developing new industry partnerships to create internship opportunities, something that was undoubtedly aided by the university's proximity to Research Triangle Park.

After the first annual meeting, the BEST Consortium identified the need to clarify definitions across the group, and decided to refer to an externship as a shadowing experience that does not involve hands-on training, and that an internship would be a more extended, hands-on experience that was typically four weeks and that rarely exceeded eight weeks. However, it is noteworthy that the internship program at the University of California, San Francisco, which was established prior to BEST in conjunction with the University of California, Davis, allowed full-time internships that were typically three months long.

At the first mention of internships, there was concern that the NIH would be opposed to using federal funds to pay for such time off. However, there has been a great deal of support from NIH institutes and program officers, and even the NSF has commenced its own granting mechanism for facilitating internships [17]. Although there isn't a blanket approval, most program officers have allowed using training grant stipends to this end as long as the internship is seen as part of student or postdoc training. If the trainee is funded by an R01 (NIH Research Project Grants), one work-around is for the institution to identify funds it can use to release the student from the lab for that time period, which can help garner faculty support for internship programs. However, federal groups (e.g., Office of Management and Budget) have advocated for faculty PIs to consider this to be a part of a trainee's professional development [18].

Many BEST programs have been committed to trying the internship route from the beginning, and they have been moving in that direction steadily such that the majority has tested this approach to varying degrees. It takes a considerable amount of time—be it for the BEST staff or for trainees—to establish each internship, and for trainees to participate. In cases where staff creates the internship, the program may benefit by establishing these relationships, which they may then find to be very valuable to help with other aspects of their program (e.g., career panels).

The experience of UNC's program describes an approach in which the use of internships has been quite extensive and successful, with about 30 students involved annually and over 100 placed to date. Although UNC's internship program offers the experience to both graduate students and postdocs, there were sometimes complications with postdocs who successfully completed an internship. Due to the nature of postdoc employment, in some instances there was a danger of postdocs receiving job offers and leaving their labs with partially completed projects to join the company where they interned, leading the university to consider limiting the offer of internships to graduate students only. These cases of dissatisfaction were extremely rare (<10%), as most PIs, internship hosts, and trainees expressed positive experiences.

Internships and externships are of course not the only form of experiential learning [19]. Two upcoming publications from the BEST programs define the range of hands-on opportunities that can yield meaningful experiences and skills developed without the longer time commitments of the above examples [20,21].

Advisory committees at individual sites

One feature of the original applications for NIH support was the requirement for identification of an external advisory committee to provide overall program guidance at each institution. At early stages in the BEST experiment, most of the new programs formed such committees and incorporated their ideas into the available offerings. However, with a couple of notable exceptions, the BEST programs have elected not to expend a substantial effort in subsequent formal interactions with their external advisors, and some have even fallen into disuse; in at least one case, however, the advisory committee turned out to be a valuable source of well-informed seminar speakers and participants in other program activities.

The reasons why most programs have not availed themselves extensively of this resource vary. Some found it difficult to get their external advisors, who have full-time jobs, to come to on-site meetings. Others did not find the meetings particularly helpful. A possible explanation is that this tends to be uncharted territory, and that the largely self-trained staff has become the expert in this very new field.

The external program consultants

In addition to the advisory committees at individual sites described above, all NIH Common Fund programs have a group of external program consultants to provide opinions to the NIH on how the programs are running and whether they are meeting their goals and milestones. These individuals are invited by the NIH staff to work on the program, and their opinions are provided on an individual basis—they are not a voting panel. These individuals attended most of the face-to-face meetings and many of the monthly conference calls. They shared their individual opinions and helped build the national consortium.

The individual consultants were not directly engaged in the day-to-day activities of the individual programs, although individual committee members made a significant number of on-site visits early in the award. They oftentimes emphasized the longer and broader perspective to many of the programs, which, of course, were preoccupied with the immediacy of establishing and implementing the many new offerings for their trainees. Many of their suggestions for data sharing and publication are only now becoming a reality as the NIH support for the programs is ending, which makes sense considering that the programs are engaged in what is essentially an extensive, long-term experiment that is only now beginning to yield results.

Lessons learned

The overall BEST experiment at the 17 institutions has been deemed to be success; individual reports, some of which are detailed in this book, exemplify the many criteria that have been applied to the evaluation and estimation of "success," although, in the long run, success will be measured in terms of the careers satisfaction these students and postdocs develop. There have been multiple internal surveys at all institutions, and, as is described in the chapters that follow, the response from the trainees has been overwhelmingly positive and supportive.

The success of the BEST experiments has been a consequence of at least three factors. First, there was clearly a huge need—and indeed, demand—for this kind of programming from both graduate students and postdocs. At the start of BEST, there was a profound inability to supply career development information for biomedical Ph.D.'s at many medical schools. The second factor, which was paramount, reflects the fact that the resources provided by the grant allowed for the hiring of staff that was utterly dedicated to building expertise in the career

development of biomedical trainees. They were familiar with the concerns of the trainees and found interactions easy to establish with them, often due to the fact that many of the staff were recent doctoral and postdoctoral graduates themselves. Finally, the resources made available by the NIH awards were finite, and so the staff involved in creating and implementing the content was also quite small. This meant that although the groups were modest in size (often individuals or pairs), they were free to test approaches quickly, to modify as needed, and to react promptly to perceived needs and concerns without having to go through complex university administrative structures even in cases where they supported over 1,000 trainees. The BEST program simply would not have worked without the hugely dedicated staff at the 18 institutions [22]; all of whom were incredibly hard-working, had many roles, and successfully fulfilled them all.

These last points bear emphasizing for the benefit of any school that would like to reproduce some of the models of the BEST experiments. For a successful program, it is not necessary to identify a national thought leader in the field; rather, it is preferable to identify a local individual who understands many of the concerns of the trainees, and to give this person full and unqualified support. Every BEST program had at least one faculty or staff member who had the vision, leadership skills, and commitment to drive the local endeavor forward.

For input about a wide range of careers, almost all programs quickly turned to dedicated local homegrown talent: alumni. Although alumni tend to be scattered all over the country, a significant number remain in the general area of each school. In addition, alumni have an understanding of the pressures on the trainees in their specific institutional context and can give advice relevant to the local campus culture. They are more likely to feel a commitment to current trainees at the university and therefore be willing to return and report on their own experiences. The alumni, who are viewed by trainees as highly approachable, have typically moved into an incredibly diverse set of careers, reflective of the options available to current trainees. With the pedagogical expertise and structure provided by career and professional development staff and the insight provided by alumni, you can soon have a course, a job simulation, or a seminar series. Any school interested in starting their own BEST experiment will find that the alumni are an incredibly effective and powerful force (with the added benefit of being non-resource intensive), and that relying on the alumni network is a strategy that has continued to be successful year after year. This approach has the additional value of helping each institution maintain fruitful interactions with its alumni.

The trainee experience

Individual programs have surveyed their trainees to assess the perceived value of the BEST experiences, which has led to many discussions of dosage and of how much exposure counts as a productive experience. When the original awards were made, the NIH contracted with an evaluation company, Windrose, LLC, to run consortium-wide surveys on participation in and satisfaction with the BEST programs, and coordinate the collection of related data tables submitted by each of the institutions, tracking items such as activities delivered, program participation and demographics, and time to degree.

Windrose surveyed the trainees at the time of recruitment into the BEST program and throughout the educational period, but its evaluation is designed to continue for at least another decade despite the fact that the NIH did not commit funds for that long. The requirement for the cross-site evaluation was clearly spelled out in the original RFAs for the awards, but the awardees were not prepared for the pressure of this cross-site evaluation. The time demands imposed on them to organize the data on their activities and participants turned out to be fairly extensive, and it possibly interfered with other projects planned by many of the groups. Nonetheless, the NIH data collection is beginning to produce results.

One of the easiest inferences to draw from the NIH cross-site data is about the trainee perception of how much they feel they are benefitting from their respective BEST programs/activities. The NIH data can also be used to determine whether trainees believe that their BEST exposure has been helpful as they begin to explore career possibilities. Initial data suggest that the majority of students — at all institutions - find the programming helpful. The NIH will report on these data separately, and it is currently working on a portal where all the de-identified data will be available to anyone who wishes to analyze them.

Various institutional assessments on the effect of career development activities on trainee outcomes indicate that individuals who chose tenure-track jobs spent proportionately less time in the broader range of career programming than those who looked towards other careers. It is imperative to caution against the interpretation that faculty positions require less support in this area. It is quite likely that trainees aiming for faculty positions are very selective in the sessions they attend. For example, interviewing and maintaining proper lab organization are important skills for them, possibly more so than exploring a wide range of career outcomes since they have already narrowed down their goals. In addition, their advisor may be able to provide them with much of the needed career advice for the tenure track. It is worth stressing that the BEST programs fully support trainees who wish to prepare for an academic career. Indeed, as alluded to above, all BEST programs have many enrichment activities that are focused directly on preparation for an academic career.

Although the above discussion has been very graduate student-centric, similar approaches work well for postdoc exposure to career development. However, the nature of the career development experiences has to reflect the fact that the average time spent in a postdoc is shorter—an average of 2.6 years for a single postdoc and an average of 3.6 years for the total postdoc training experience [23]—and so they have to begin their exposure to career options as early as possible. Other than that, postdocs in BEST institutions have mostly been able to pick from the same menu of training activities as graduate students.

Working together

The institutions in the BEST Consortium always assumed they would interact with each other in order to facilitate some of their activities, and, in fact, the nature of their interactions has been extensive at a variety of levels. Systematic communications efforts have centered primarily on monthly conference calls and annual meetings. Individual programs have emphasized how useful these meetings have been in terms of finding out how things were progressing at other

schools and in learning how to implement new activities. A number of collaborative projects, including team-written manuscripts, have organically developed as a result of these interactions, which has allowed programs to develop intra-institutional connections and reinforce the consortium's collaborative spirit.

Committees

The formation of a number of committees has fostered widespread formal (and informal) intra-institutional communication. The committee that highlights this the most is the Project Coordination Committee (PCC), the objective of which is to disseminate project ideas and facilitate collaborations and which is composed of one person from each institution. When new joint project ideas (e.g., workshop, manuscript, or presentation proposals) are formed, the initiators of that idea complete a short Project Notification Form, which is sent to each PCC member. Although the NIH BEST funding ended in September 2019, the PCC will continue to function. This may prove a powerful tool into the future, allowing for continued interactions among all BEST institutions.

Another source of interactive activity was the annual meeting planning and steering committees. The membership in these committees was shared among all programs over the years. Sharing responsibility ensured that all the member institutions had a voice in the consortium and fostered creativity from year to year. Typically, the annual meeting goal was to highlight ideas from within, but it also strove to help the program members keep up to date with the breadth of career training activities across the country.

Outreach

The BEST programs have played a role in providing outreach to other graduate programs by offering detailed information about their experiences. This kind of outreach has involved a great deal of interaction between the BEST administrators in terms of the creation of materials that describe BEST activities. One of the major efforts in this direction involved a detailed presentation that reflected all 17 programs, which was to be presented as a full-day activity immediately prior to the 2017 GREAT Group meeting in Orlando, FL, and which was created through extensive collaboration by the entire group. The Burroughs Wellcome Fund and the NIH provided financial support to encourage GREAT Group meeting participants to arrive before the regular meeting and attend the presentation by BEST. Preparing for this meeting was incredibly valuable because it allowed the group to take stock and review the program *in toto*. Unfortunately, the presentation was not fully delivered due to the unwelcome arrival of Hurricane Irma. Nonetheless, the content from the 2017 GREAT Group meeting was summarized and still resides on the NIH BEST website [24]; readers of this book are encourage to visit the site.

The following year, the BEST Consortium gave a detailed data collection and career outcomes analysis presentation at the 2018 GREAT Group meeting. This presentation featured a very productive interaction with the Oregon Health & Science University graduate programs, which has been collecting similar data on career outcomes independently of the BEST Consortium. These interactions set the stage for the participation of groups from outside of the consortium at the final BEST meeting in 2018.

Another example of collaboration with groups not in the consortium is the BEST Beginning Enhancement Track (BET), which is an initiative supported by an NSF program that supports diversity and inclusion. This collaborative effort draws upon the expertise of five research-intensive institutions from across the country (Boston University, Cornell University, University of Colorado Denver, UNC, Wayne State University). The goal of the program is to focus on career exploration to engage traditionally underrepresented undergraduate students who are studying science.

Developing a career taxonomy

An important area of collaboration has been the development of a unified career taxonomy. Initially, the focus was on very broad career trajectories—academia, industry, government, non-traditional careers, and careers out of science—but it rapidly became apparent to all the schools that there is a very wide range of subdivisions in these consolidated groups [25]. For instance, academia can accommodate a wide range of possible career variations: teaching at a two-year or four-year college, becoming a research or tenure-track faculty, working within high-tech cores, or working as academic administrative support. As individual programs identified the specific outcomes of their trainees, it became evident that comparison between schools was very difficult without a commonly defined and agreed-upon taxonomy. This led to a major commitment by the consortium members to develop a common language to describe the numerous and complex outcomes for highly trained individuals within the biomedical research Ph.D. space. The structure of the outcomes taxonomy has recently been used as the basis for a ten-school group of institutions, the Coalition for Next Generation Life Science (NGLS), to indicate to the public the nature of their outcomes in the biomedical science research arena [26]. Certainly, a common taxonomy throughout all graduate programs will be of immense value as it will truly allow for comparison in terms of outcomes between all U.S. programs involved in biomedical research training.

A major benefit of the BEST program has been that it has, in part independently and in part collectively, found a mechanism to obtain reliable information on student/postdoc career outcomes in a way that is both rapid and inexpensive and that can be used directly for a taxonomy outcomes compilation. The process involves careful training of an individual on searching LinkedIn and other Internet sources (e.g., Facebook, company websites, etc.). Most uncertainties can be resolved by direct contact with a given individual, but such direct contact is needed only infrequently. However, a vast yield of outcome information is relatively useless unless one can truly compare one set of findings with another.

The taxonomy group worked hard to engineer strategies to ensure reproducibility in identifying specific outcomes. The revised taxonomy was tested by one set of individuals, and retested, using the same data set, with a different set of coders to determine the veracity of the first set. Initially, the percentage of the data that was corroborated was less than desirable, necessitating a (successful) revision of the overall taxonomy so that numerous independent reviewers could code the same outcomes with >95% reproducibility [27]. Without extensive interactions, this kind of precision and cross-institutional collaboration would not have been easily attained.

As of late 2018, many BEST programs commenced their adoption of the broad taxonomic structure; however, this is a work in progress and continues to evolve as it is adopted in part or in whole by additional institutions. At the time of writing, as indicated above, a number of NGLS institutions have agreed to increase career outcome transparency by making a commitment to publish their trainee career outcomes data online [28].

Career outcomes of trainees at BEST institutions

Although the initial NIH data indicated that trainees value their experiences in BEST, the key question ultimately is: what are the actual career outcomes across the BEST schools? As many trainees are still in training, it is too soon to have concrete data on career outcomes. Consequently, comparative data are not available, but it is possible to reflect on the ongoing use of the taxonomy. One can visit the consortium website and find a list of schools that publish their overall training information [29] (e.g., demographics, time to degree, degree program) and career outcomes on their website, though the style of the presentation varies from school to school as the data are largely part of the recruiting strategies of the individual programs; at this time, at least seven institutions have committed fully to using the developed taxonomy. For this reason, the programs have not felt the need to publish aggregated consortium data. Nonetheless, it is possible to visit these sites and come up with roughly comparable analyses and to figure out what the overall outcomes are in terms of chosen trainee career paths, which ultimately benefits future doctoral students by allowing them to make informed choices about where to attend based on comparable outcomes of interest to them.

The BEST schools that publish their outcomes online report varied results. For example, institutions report that between 35% to 55% of their newly minted Ph.D.

trainees move into postdoctoral positions after graduation. This wide range demonstrates the need for schools to publish their outcomes data online, giving future graduate student more power when making career decisions.

Nationally, approximately 50% of new biomedical Ph.D. holders stayed in academia in 2013, according to a report [30]; as stated above, however, academia is a very broad category that covers traditional tenure-track, non-tenure track research, research staff, administrative, and policy positions.

The percentage of biomedical postdocs who obtain tenure-track positions each year is approximately 30% [30]; and only half of those are R1 tenure-track positions [31]. The remaining 70% of individuals completing postdoctoral positions find careers in several other employment sectors (government, for-profit, non-profit, or other) that benefit from their advanced training in biomedical research.

Following students and postdocs as their academic and subsequent careers unfold involves a commitment to fairly long-time horizons. This has meant that long-term studies between individual institutions are only now beginning.

Recording extensive data and overall outcomes reflects one of the earliest goals of the consortium as a whole: to publish up-to-date information about the outcomes of the individual programs on a public website. The BEST group assumed that this tactic would play a key role in recruiting the highest-caliber students to their graduate programs, and called it "truth in advertising." The approach has now been copied by many other schools involved in graduate education, which is surely a measure of one of the impacts that BEST has had on graduate education across the country.

Trainee diversity

The NIH has expended a great deal of effort in the last decade or so to diversify the biomedical research workforce. Although progress has been slow, there is no doubt that the NIH is meeting its goals. The number of underrepresented minority (URM) students graduating with a Ph.D. in the biomedical research arena now reflects the proportion of URM students that graduate with a B.S. degree in the United States [32]. Thus, it is appropriate to ask about the level of participation of various demographic groups in BEST activities, and particularly, whether all groups have equal access to the new and comprehensive services supported by the BEST programs. This is a question the BEST Consortium members are interested in testing, and have developed a collaborative group to combine data to assess professional development participation and effectiveness for trainees from intersecting social identities.

The initial funding opportunity announcement had no stipulation for studying this issue; however, a recent informal survey of the individual programs asked about participation and inclusion of all demographic groups. Although there was no specific imperative to collect demographic data as a part of participation in BEST, all of the programs responded that as best as they could determine, participation by all demographic groups more or less represented the distribution seen in their graduate and postdoctoral programs. Certainly, there is no evidence of unequal access or attendance at the current time. Thus, in absence of evidence to the contrary, it seems likely that the services provided by the BEST initiative are about as attractive to trainees from diverse backgrounds.

xxviii   Origin of BEST, and how 17 very different programs created 17 related approaches

And it wasnt all sunshine...

Throughout the five-year BEST journey, the NIH stressed that the whole enterprise was an experiment. As such, although they hoped that many of the ideas tested would be effective and would work out well, the NIH sponsors encouraged the BEST groups to note the aspects and approaches that failed or that did not work well. Although the students and postdocs were highly supportive of the significant and sincere institutional efforts set forth by BEST, it is worth mentioning that some approaches and practices did not pan out as expected. The following are examples that come from each of the institutions featured in this book.

Boston University based its plan of attack on analyzing nationwide job listings to determine real-time information about the job market (e.g., job titles, employers, skills, etc.) and used this knowledge to guide programming. BU initially collaborated with a local non-profit already licensing the software used to generate the workforce data, but it was recognized early on that BU needed direct access to the software.

Cornell University found that attendance at some events was low and concluded they needed a more targeted outreach. They worked strategically with their advisory board (comprised exclusively of graduate students and postdocs), individual students, and student organizations to create interest and to help with event planning, and saw a subsequent rise in attendance.

The Atlanta BEST Program at Emory University and the Georgia Institute of Technology recognized several approaches that did not work out well for their program, which was based on a cohort model. First, they learned early that cookie-cutter programming was bound to fail. For example, they initially required all BEST trainees to undergo exploration of specific career tracks,

such as technology transfer, but not all trainees were invested. Second, stand-alone workshops typically failed to move trainees forward and were therefore of limited value; instead, they opted for programs that built skills over time, enabling trainees to undertake full exploration of specific subjects and thereby gaining confidence. Finally, internships/externships do not always work and are certainly not required for trainees to recognize that a particular career track is not for them.

Michigan State University found that if the BEST team organized social activities in an effort to build "team cohesion" for its cohorts of BEST trainees, they were not well subscribed. However, if the students took over the organization of these activities then they became much more popular.

New York University tried an interesting approach to teaching communication skills that turned out to be expensive and too intense for the majority of trainees. The approach used individual acting and theater techniques, meeting for three hours per week with the idea that the participants would end up doing a TED-style talk at the end of the semester. Although this was a creative plan and those who completed the course found it extremely beneficial, NYU found that it struggled to keep participants from dropping out and rarely got above a 50% completion rate. The university found that it was too resource intensive for the benefit of only a small number of trainees and modified the course to be able to impact a larger number of trainees. NYU turned this course into a single workshop in which participants work on communication skills without having to commit an entire semester, which has allowed it to triple the number of participants benefitted at less than half the initial cost.

University of Rochester had trouble with peer-facilitated events, which had a low turnout and sustainability; it appeared that graduate students and postdocs were looking for guidance from people that had successfully transitioned into science or science-related careers. As a result, Rochester is currently monitoring two graduate student groups from a Blue Sky Visioning workshop led by Rebecca Layton, Ph.D., as they continue over the course of one year. These events have a great chance of surviving because Layton provided the student groups with a structure for running their meetings early on and will be following up with them at 6 and 12 months, but it is as yet unclear if the provided structure will be enough to keep the graduate students accountable.

Rutgers University faced challenges due to the spread of its four campuses across New Brunswick, NJ. It effectively overcame the issue by offering transportation between campuses and being cognizant of having events at each one.

The University of California, Davis, tried offering a self-directed track as a less time-intensive alternative to the cohort model of their program. The track condensed their career exploration course into a one-day workshop featuring course content highlights. The goal was to serve more trainees than could be accommodated in the cohorts, and to enable participants to get the basics in career exploration and preparation without making a quarter-long commitment. Although a large number of graduate students and postdocs signed up to participate in the self-directed track, very few followed through with attending the one-day workshop, and most of those who attended did not stay engaged with the program. UC Davis tried breaking up the workshop into two half days and offering it on different days of the week, to no avail. Finally, the program decided to scratch the self-directed

option, and instead increased the number of cohorts offered from one to three per year.

The University of California, Irvine, tried a well-intentioned approach. There is a professional organization in Southern California that brings in start-ups and business people to talk about trends in medical devices. The BEST group thought that the students would benefit from attending these events by mingling, as the organization was willing to host a limited number of trainees. However, some students reported feeling like fish out of the water as they didn't feel like they had the preparation needed to interact and converse.

An innovative concept from the University of California, San Francisco, was MIND-Bank, which formalized a range of possible mentors for the trainees to contact. Unfortunately, the students preferred a much less formal approach and used their own networks as starting points, and, as a result, this sophisticated tool was much underutilized.

University of Chicago's BEST program, myCHOICE, has closely monitored and analyzed event attendance, revealing a steady 20% participation rate from the postdoctoral community but a shift in the number of graduate students who participate, from 38% during the first full year of programming to 25% in the program's fourth year. Closer analyses of the data reveal what we believe to be a "pent-up demand" phenomenon in which myCHOICE was novel to every community member in our first year but which created a population of older graduate students who no longer needed the same content—namely, a basic exposure to the diverse career paths taken by Ph.D.-trained scientists—by their fourth year. Longitudinal analyses have corroborated these observations: the seminar series is primarily attended by early-stage trainees or those who are new to campus, while later-stage trainees participate in activities that reflect a pre-existing familiarity with their career options, including internships, specialized mini-courses, and events that involve networking with alumni. Similarly, decreased demand for specific types of annual mini-courses provided information on trainee turnover: at the University of Chicago, there are insufficient new trainees to warrant repeating events more frequently than at 16-month intervals. Although the normalization of attendance according to stage of training was expected, the decrease in attendance numbers has required us to strategically plan our events on a longer timescale than was originally intended and to modify our attendance expectations accordingly.

CU Denver|Anschutz Medical Campus's BESST Program set up monthly peer meetings with students and postdocs in the biomedical sciences as a forum to build a community among peers from different programs, and to discuss needs and interests regarding career development. Originally, the meetings were called and guided by the BESST Program Director, but students and postdocs later volunteered to host the meetings, with lunch provided by the BESST Program. Concerns that the needs of postdocs might be too different from those of students led the program leadership to meet with trainees in separate groups, but the separation did not improve participation. Meanwhile, other, less formal trainee-initiated opportunities to meet and collaborate around a common goal had been developed on campus that also fulfilled the function of community building, and the peer meetings were therefore disbanded.

When the University of Massachusetts Medical School obtained its BEST award, it had only held a limited number of career development events on its campus. One early element integrated into the curriculum was a course focused on IDP and career awareness for third-year students. A few years later, after implementing BEST-led changes in the earlier stages of the curriculum, UMassMed found that later cohorts of students were more comfortable with the basics of career preparation by their third year. As a result, the university adapted the third-year course to address higher-level skills in career planning and development.

The University of North Carolina found that IDP workshops were initially unpopular because students were intimidated by the focus on formal conversations with their PIs around the IDP tool and their career planning. Instead, UNC rebranded the workshops, and it now introduces the idea in two phases made up of a career exploration and a career preparation component. It is much better attended and is more effective, yet it remains consistent with the concept behind the IDP and still encourages students to have career conversations with their mentors. The rebranded workshop focus has shifted to help trainees identify action steps to foster their professional development planning rather than rely on external sources.

Vanderbilt University initially required that the graduate students in the umbrella program attend several BEST events. It found, however, that making attendance mandatory decreased attendance. When it removed the requirement and made attendance optional, attendance increased, suggesting that it was better to let students follow their own interests.

One challenge that Virginia Tech faced was the misconception that BEST programming was only for trainees who wanted to pursue non-academic careers, despite messaging that the aim was to provide programming important for biomedical trainees regardless of their ultimate career path. One mechanism for addressing this concern and for better contextualizing the information was to incorporate content into the core graduate coursework. For example, Virginia Tech offers a module on entrepreneurship, where teams of trainees from the biomedical sciences, engineering, and business are mentored by professionals in the identification and development of intellectual property for a business pitch competition. The module was integrated into existing for-credit biomedical coursework that instructs on the fundamentals of clinical and translational science. Thus, regardless of career goals, all students developed a better understanding of how professionals in academia, intellectual property, regulatory, venture capital, and industry contribute toward translating discoveries in their scientific field, in addition to developing valuable communication, budgeting, and team management skills.

Wayne State University had a module called Community Engagement, as it was identified as a desired topic through internal surveys, but the turnout was poor and it was hard to get people to sign up. WSU figured out that rather than teaching community engagement as a stand-alone module, it could incorporate the lessons into other modules, and that approach was successful.

The future: Lo que será, será

All the institutions undertook the BEST responsibilities knowing that no matter what interventions were proposed, nor how successful they were, the NIH funding would not extend beyond five years. At the conclusion of the grant period, institutions needed to identify other sources of support if they wished to continue their endeavors. Initially, programs did not think about the transition to other funding sources, but as time went on, this became a bigger concern, especially as they approached their final year. Fortunately, most of the programs were able to convince their senior administration that this kind of career training support was the right thing to do, and that BEST activities were incredibly valuable for students and postdocs and made for happier and more productive trainees. Consequently, most of the programs will continue operating more or less as they did during the NIH grant period. Several of the programs will be relocated and housed within a graduate school or moved into a broader career development office.

It is fair to say that the institutions that participated in the BEST experiment will continue to reap the benefits of the collective creative input of the last five years going forward.

Overview

In the following chapters, the reader can appreciate how the individual institutions addressed their participation in BEST. This book attempts to provide sufficient information for others to follow the group's guidance, but the readers are encouraged to reach out to particular institutions to learn directly from the sources, who would be happy to share insights that go beyond the content of this book. Interested parties may wish to focus on the approaches used by institutions with which they have much in common, a tactic that should not be difficult as there is a wide diversity of types of institutions. In addition, although there are many stories of success, they have been achieved in numerous different ways.

The programs have leveraged the talents immediately available in their faculty and staff, and have created structures that have been generally successful. In most cases, the dedication of the staff has allowed them to turn an initial lack of expertise in this area into self-grown strength in support of their trainees. The overall cost of these activities is surprisingly small, and, at an annual expense that is less than that of running a single, moderately equipped lab, the BEST programs have found mechanisms to support hundreds of trainees (or more) at each institution. Many schools reported that one of their greatest successes was a culture shift in the university faculty towards acceptance of non-academic career pathways for their trainees.

The BEST institutions that contributed to this book hope that readers will appreciate the trials and tribulations of setting up new systems, and the enjoyment that stems from helping so many people who are eager to find a rewarding career that reflects their commitment to science.

The following chapters will provide an overview of different approaches. Each chapter will introduce successful strategies and will offer guidance on how to implement similar activities at the reader's own institution. Readers who seek more detail can reach out to the specific institutions.

Acknowledgments

All BEST programs owe a major debt to the NIH Common Fund and their staff for their euthusiastic support. On behaf of all the BEST programs, we want to acknowledge the incredible support and the constant help and encouragement we received from Patricia Labosky and Rebecca Lenzi, who were an invaluable inspiration throughout. Labosky and Lenzi both work in the Common Fund within the Office of the Director at the NIH.

References

[1] Graduate Career Consortium website. Available from: www.gradcareerconsortium.org/about.php.

[2] Careers '95: the future of the Ph.D. Science 270 (5233), 1995. Available from: http://science.sciencemag.org/content/270/5233.

[3] National Postdoc Association website. Available from: www.nationalpostdoc.org/.

[4] Jaschik, S., 2005. Postdoc (partial) satisfaction. Inside Higher Ed. Available from: www.insidehighered.com/news/2005/04/06/postdoc-partial-satisfaction.

[5] Powell, K., 2010. The postdoc experience: high expectations, grounded in reality. Science. Available from: www.sciencemag.org/features/2012/08/postdoc-experience-high-expectations-grounded-reality.

[6] Cyranoski, D., Gilbert, N., Ledford, H., Nayar, A., Yahia, M., 2011. Education: the PhD factory. Nature 472 (7343), 276–279.

[7] Iasevoli, B., 2015. A glut of Ph.D.s means long odds of getting jobs. nprEd. Available from: www.npr.org/sections/ed/2015/02/27/388443923/a-glut-of-ph-d-s-means-long-odds-of-getting-jobs.

[8] Meyers, L., Brown, A.M., Moneta-Koehler, L., Chalkley, R., 2018. Survey of checkpoints along the pathway to diverse biomedical research faculty. PLoS One 13 (1), e0190606.

[9] Garrison, H., Justement, L., Gerbi, S., 2016. Biomedical science postdocs: an end to the era of expansion. The FASEB Journal 30, 41–44.

[10] National Institutes of Health, 2012. Biomedical research workforce working group report, Bethesda, MD. Available from: https://acd.od.nih.gov/documents/reports/Biomedical_research_wgreport.pdf.

[11] National Institutes of Health; 2012.

[12] National Institutes of Health, 2013. NIH director's biomedical research workforce innovation award: Broadening Experiences in Scientific Training (BEST) (DP7). Available from: https://grants.nih.gov/grants/guide/rfa-files/rfa-rm-12-022.html.

[13] Davis, G., 2005. Doctors without orders. American Scientist 93 (3 Suppl.). Available from: http://postdoc.sigmaxi.org/results/.

[14] Watts, S.W., Chatterjee, D., Rojewski, J.W., Shoshkes Reiss, C., Baas, T., Gould, K.L., Brown, A.M., Chalkley, R., Brandt, P., Wefes, I., Hyman, L., Ford, J.K., 2019. Faculty perceptions and knowledge of career development of trainees in biomedical science: What do we (think we) know? PLoS One 14 (1), e0210189. Available from: http://dx.doi.org/10.1371/journal.pone.0210189.

[15] In preparation.

[16] Watts; 2019.

[17] Graduate research internship program, NSF 18-069. Available from: https://nsf.gov/pubs/2018/nsf18069/nsf18069.jsp.

[18] Office of Management and Budget 2, Code of Federal Regulations 200.

[19] National Society for Experiential Education, 1998. 8 principles of good practice for experiential learning activities. Available from: www.nsee.org/8-principles.

[20] Wart A, O'Brien T, Brady AE, Varvayanis S, Alder J, Greenier J, Layton RL, Stayart CA, Wefes I. Experiential learning for career development — program development and design [In preparation].

[21] Chatterjee, D., Ford, J.K., Rojewski, J.W., Watts, S.W., 2019. Exploring the impact of formal internships on biomedical graduate and post-graduate careers: an interview study. Life Sciences Education 18 (2), 1—13.

[22] Although 17 awards were made, the Georgia Institute of Technology was a sub-awardee of Emory University.

[23] Andalib, M., Ghaffarzadegan, N., Larson, R., 2018. The postdoc queue: a labour force in waiting. Systems Research and Behavioral Science 35 (6), 675—686.

[24] Proceedings of the BEST practices workshop, 2017. Available from: http://www.nihbest.org/2017best-practices-workshop/.

[25] Mathur, A., Brandt, P., Chalkley, R., Daniel, L., Labosky, P., Stayart, C.A., Meyers, F., 2018. Evolution of a functional taxonomy of career pathways for biomedical trainees. Journal of Clinical and Translational Science 2 (2), 63—65.

[26] Coalition for next generation life science. Available from: http://nglscoalition.org/.

[27] Stayart, C.A., Brandt, P., Brown, A.M., Hutto, T., Layton, R.L., Petrie, K., Flores-Kim, E., Peña, C., Fuhrmann, C., Monsalve, G., 2018. Applying inter-rater reliability to improve consistency in classifying PhD career outcomes. bioRxiv. Available from: http://dx.doi.org/10.1101/370783.

[28] Coalition for next generation life science; n.d.

[29] Tracking participation and evaluating outcomes, n.d. Available from: http://www.nihbest.org/build-career-development-program/tracking-participation-evaluating-outcomes/.

[30] Andalib; 2018.

[31] Biomedical research workforce working group report; 2012.

[32] Meyers; 2018.

BU's BEST: Using biomedical workforce data to inform curriculum and influence career exploration

*Amanda Florence Bolgioni[a], Chelsea R. Barbercheck[a,f],
Sarah Chobot Hokanson[b], M. Isabel Dominguez[c],
Jean L. Spencer[d], Linda E. Hyman[e], Barbara M. Schreiber[d]*

[a]BU's BEST, Department of Medical Sciences & Education, Boston University School of
Medicine, Boston, MA, United States; [b]Office of the Provost, Boston University, Boston, MA,
United States; [c]Department of Medicine, Boston University School of Medicine, Boston, MA,
United States; [d]Department of Biochemistry, Boston University School of Medicine, Boston,
MA, United States; [e]Graduate Medical Sciences, Boston University School of Medicine, Boston,
MA, United States; Marine Biological Laboratory, Woods Hole, MA, United States; [f]Great
Plains IDEA, Kansas State University, Manhattan, KS, United States

Introduction

Graduate training in biomedical research has long been directed toward careers in academic research, but a transformation in the way we train scientists is essential to fulfill the needs of the current and future biomedical workforces. Only with trainee exposure and training for a multitude of different career options can we ensure that the workforce will be supplied with well-trained personnel. With this goal in mind and with funding from the National Institutes of Health (NIH), we launched Boston University's Broadening Experience in Scientific Training (BU's BEST) program to "re-engineer" biomedical graduate training. The needs of employers (e.g., biotechnology industry, academic institutions, government, law firms) inform the development of our curriculum, which enables trainees access to up-to-date information on jobs and the skills required to secure them.

This chapter will describe our approach to developing programming and guiding informed career choices. Modeled on a classic "feedback loop," workforce needs are

BEST: Implementing Career Development Activities for Biomedical Research Trainees
https://doi.org/10.1016/B978-0-12-820759-8.00001-2

determined to enable the trainees to consider relevant career options and the faculty and staff to identify potential skill gaps in the curriculum. The gaps can then be mapped onto the trainee curriculum, thereby supplying the workforce with individuals trained to meet defined workforce requirements (Fig. 1.1). The Commonwealth of Massachusetts and the greater Boston metropolitan area in particular are sensitive to the needs of both the "trainee" and the "workforce" sides of the equation. With more than 100 universities and colleges and more than 800 companies representing all facets of the biotechnology industry in Massachusetts, BU is well positioned to execute these efforts.

BU's BEST program engages trainees from all BU schools and colleges who are involved in biomedical research, which includes ~1,000 doctoral and postdoctoral trainees who are eligible to participate in our programming. By taking the multi-pronged approach of offering workshops, panel discussions, site visits, internships, and credit-bearing courses, BU's BEST provides biomedical research trainees with opportunities to explore and gain experience in achieving their individual career goals, while at the same time providing the workforce with employees trained to meet its needs.

Here, we describe the methodology we use to determine workforce needs as well as some examples of how these data are used. Understanding that all efforts to develop the programming described can be expensive, we have suggested some cost-saving options. Along with the other NIH BEST awardees, the evaluation of the effectiveness of BU's BEST is implemented at the national level. Additional assessment is accomplished at the local level to evaluate both the success of the data collection and the programming opportunities offered by BU's BEST.

Generating workforce data

As this program was conceived with the intent of "re-engineering the training pipeline," the analysis of the job market is designed to help trainees consider job options, prepare themselves to secure those jobs, and help the university faculty and staff take these aspects into consideration when making curricular decisions for career development programming. To analyze the biomedical job market, we use a software tool called Labor Insight™ that is offered by Burning Glass Technologies [1]. The search criteria are set by the user so as to query the job market in different sectors and regions or by the required education levels, for example. Millions of job postings are scanned to generate the workforce data. We

FIG. 1.1   Feedback loop. BU's BEST takes an evidence-based approach to determining workforce needs.

categorized our searches into the following broad biomedical career areas: business/administration, communication, law, policy, research (academic and industry), and teaching. Figs. 1.2 and 1.3 show examples of data generated for research jobs in the biomedical industry in 2017, including regional demand for the jobs (Fig. 1.2) and the skills required to secure the jobs (Fig. 1.3).

Limited resources?

Although the cost of the software license can pose a barrier, the use of publicly available workforce data, such as from the Bureau of Labor Statistics or the United States Department of Labor, can be considered. State-specific economic workforce efforts (e.g., MassBioEd, Delaware Bio) are very helpful as well.

The information on skills consists of specialized or hard skills (Fig. 1.3A), baseline or so-called soft skills such as communication or leadership skills (Fig. 1.3B), and computer skills (Fig. 1.3C). Additional information that can be aggregated includes top job titles and employers. Moreover, trend analyses can be determined to indicate changes in the demand for a particular job or career path. The reports are disseminated through the BU's BEST website, making them available for all trainees, faculty, and staff [2].

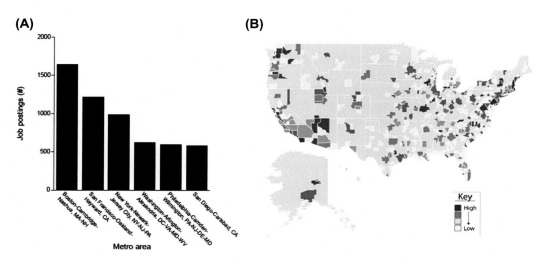

FIG. 1.2 Top areas for biomedical industry research jobs by metropolitan statistical area (MSA). (A) The regions with the most biomedical industry research job posts in 2017 are depicted. (B) The nationwide distribution of job postings is shown. The scale indicates where there are more jobs (dark blue) or fewer jobs (white). *Data were generated using Burning Glass Technologies: Labor Insight™.*

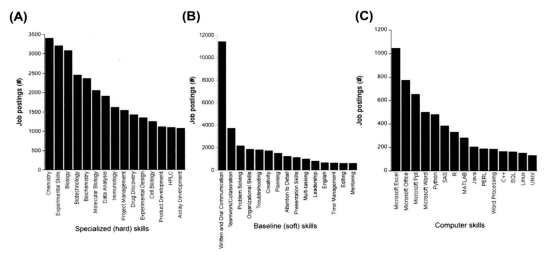

FIG. 1.3   Top skills for research in the biomedical industry. The top skills required for biomedical industry research jobs in 2017 consist of (A) specialized (hard) skills, (B) baseline (soft) skills, and (C) computer skills. *HPLC; high-performance liquid chromatography. *Data generated using Burning Glass Technologies: Labor Insight™.*

Using the workforce data

How are the workforce data used? For the trainees, the data enable the consideration of interests that can be integrated into individual development plans (IDPs) and discussions with mentors. More importantly, as trainees formulate realistic career goals, they are encouraged to consider knowledge gaps and work toward mastering needed skills.

BU's BEST faculty and staff rely on the data to facilitate our programming, which is categorized into the same broad career areas for which the search data are generated (business/administration, communication, law, policy, research, teaching). In addition to each career path, we also provide a category of general offerings that helps the trainee prepare for the job search and interview processes. Examples of our offerings are shown in Fig. 1.4.

The job market data have provided information that has resulted in various types of new workshops or even full credit-bearing courses. For example, computer skill demands were noted in various career paths, so workshops on data science were added. A full course on leadership skills was also added (see below). In addition, the identification of critical skills required to secure jobs has suggested topics to embed within existing courses to refresh coursework to reflect market needs. For instance, we added a series of R programming lessons to a required professional skill development course for Ph.D. students after noting the demand for specific computer/programming skills. We also offer one-on-one career coaching for Ph.D. students and postdoctoral researchers so they can follow up on their career planning. What follows are some detailed examples of how we have used the data to encourage our graduate students and postdoctoral researchers to prepare for the workforce.

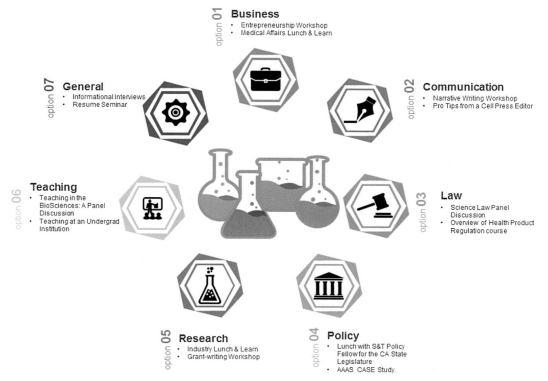

FIG. 1.4 BU's BEST Programming. BU's BEST sets up programming based on workforce needs in various career paths in the biomedical sciences. In addition to the six broad career areas, a category called "General" offers skill enhancements appropriate for all careers. Examples of offerings are listed in each category. *S&T; Science & Technology.

Encouraging trainees to write and take ownership of their career plans

First and foremost, trainees are encouraged to contemplate their career goals. Understanding the job market is certainly one important aspect to consider, but the trainees' plans must take into account personal interests and goals. Trainees enter graduate or postdoctoral training at different stages of reflection on their desired career pathways. Whereas some trainees have identified the specific career paths they would like to pursue, others have yet to decide which career areas best fit their skills, interests, and values. Generating a written career plan can help trainees reflect on career goals and, in conjunction with the workforce data, can help them determine how to best achieve those goals (e.g., they can identify skill gaps and speak with mentors about following up on career plans). In collaboration with the Professional Development & Postdoctoral Affairs Office at BU and institutions within the Center for the Integration of Research, Teaching and Learning (CIRTL) Network [3], we established a career planning workshop specifically dedicated to helping trainees develop the beginning stages of a career plan, focusing the planning phase on four main questions:

- Skills assessment: What skills do I have?
- Career aspirations: What career pathways interest me? If unsure, what do I like to do, and what do I value about my work environment?
- Desired skills: What are my goals?
- Professional development: What resources can I use to obtain my goals?

The workshop itself helps trainees identify the connections between their answers to these questions. Through the use of improvisation techniques, trainees brainstorm and practice having career-based conversations with their mentors.

Other campus units rely on a tool called myIDP for developing an IDP [4]. This free online tool allows the user to identify skills and interests and suggests scientific career paths that fit. The user is then encouraged to set timely goals for achieving aspirations.

Embedding new content into an existing professional skills course

The workforce analyses helped inform content in existing courses. For example, a course entitled "Professional Development Skills" is required for Ph.D. students in our Program in Biomedical Sciences. The students attend the class during the second semester of their first year in the program. The course objective is to extend students' education beyond the traditional biomedical course content to enable the critical development of professional skills. The course covers skills required for progress through the Ph.D. program and those that are important for specific career paths.

Course themes include (1) developing presentation skills, (2) compliance/ethics and the law, and (3) personal professional development. Although the latter theme is perhaps a misnomer because all sessions contribute to the students' personal professional development, it is highlighted by students working on an IDP and learning about a variety of career options available to those with Ph.D.s in the biomedical sciences, learning that is enabled by BU's BEST and the workforce data. In addition to a lesson on effective teaching and mentoring, there are panel discussions that feature participants working in a host of career paths. The panelists are invited to attend a lunch with some of the students as a follow-up networking event. The trainees ask the panelists about their career paths and about the skills required to secure their jobs, thereby reinforcing the workforce data generated by Labor Insight™.

L i m i t e d   r e s o u r c e s ?

Local alumni are great resources for trainees, and they are often willing to come in for a two-hour panel discussion. The "Professional Development Skills" course holds four to five panel discussions throughout the term that are open to all trainees, even those not registered in the course.

An important aspect of the personal professional development component of the class is encouraging students to seek additional activities that will contribute to their own goals and fulfillment of their careers. Students are required to identify such activities on their own, albeit with guidance and approval from the course director, and map them out early in the semester. Examples of these activities include BU's BEST offerings and other on-campus activities, such as those offered by Advance, Recruit, Retain & Organize Women in STEM (ARROWS) [5], BU's Toastmasters [6], the Center for Teaching & Learning (CTL) [7], and IS&T Research Computing Services tutorials (e.g., Excel). Additional off-campus options have included those offered by MassBioEd [8] and the CIRTL Network [9]. To fulfill the activities requirement, students can also read a book relevant to their interests in science. Some students choose to participate in volunteer activities, such as STEM education events at elementary schools. We have a program in which alumni who have agreed to serve as mentors to our students make themselves available to them. Contacting these alumni can also serve as a course-related outside activity and an important way to follow up on information gleaned from the workforce data.

The course draws on a wide variety of experts throughout BU, including faculty and staff, as well as others from outside the university. Analysis of top job titles is an influencing factor that determines which alumni are invited to participate in the panel discussions. Collectively, the course gives students an opportunity to consider career goals early in their graduate training.

Developing new coursework

Workforce data can identify curricular gaps and can catalyze the development of new courses. The biomedical workforce data consistently noted leadership and associated skills such as communication, presentation skills, teamwork/collaboration, people management, and the development of effective relationships as important for the examined sectors. As our biomedical trainees generally don't have prior opportunities to develop these skills, we designed a course entitled "Introduction to Leadership for Biomedical Education."

L i m i t e d   r e s o u r c e s ?

1. There are many free personality/traits tests available online, but make sure to properly research them and vet them before use.
2. Offer the course to master's students and staff using a tuition model as a revenue source for your institution.

3. Offer workshops instead of full courses. See Facilitator Training Opportunities at The Leadership Challenge for an example [10].

This course is specifically focused on leadership rather than on management and gives trainees a foundation for experimentation with their own leadership styles. It is designed for those who want to lead in any capacity in the biomedical sciences, including those who want to start their own laboratories in academia, pursue a position in industry, or work in a non-profit environment.

The objective is to focus on basic principles of personal and interpersonal leadership as they relate to the wide range of biomedical and health science careers. Course themes are the historical viewpoints of leadership, relationship-centered theories, self-assessment, voice, credibility, vision, emotional intelligence, communication, conflict resolution, teamwork, risk, power, and team celebrations.

To our knowledge, a textbook for leadership in the biomedical sciences does not exist. Therefore, we used the general leadership-based *The Leadership Challenge* by Kouzes and Posner, fourth edition [11], adapted it to the biomedical sciences for classroom discussions, and supplemented it with other materials such as case studies and videos from various sources.

Participation in group discussions is a major part of the final grade, and students are encouraged to think critically about innovative leadership and its application to their own leadership styles. The course culminates with a presentation in which students must demonstrate a clear understanding of the material, the ability to evaluate leadership styles (including their own), and the ability to apply the concepts learned in class.

Experiential learning: Internships and site visits

Equipped with the knowledge from IDPs, workshops, and courses, biomedical trainees can gain additional exposure to different career paths through internships and site visits, further developing their skills and confirming what they learned on campus about workforce needs.

Internships in different career tracks are offered in order to allow trainees to focus on a field of interest. Although the time commitment varies depending upon the internship, 10—15 hours per week for three to six months is typical for both on- and off-campus internships. Examples of the on-campus internships that are offered include working with Institutional Review Board personnel, grants administrators, faculty on lecture/exam development, and master's students completing their theses. Examples of off-campus internships include working in a consulting or venture capital firm and doing communication/marketing work at a laboratory incubator space.

L i m i t e d   r e s o u r c e s ?

1. Advertise internships organized by others. Many companies, for example, offer these opportunities.
2. Work with your university's career center, as they often serve as a repository for internship activities for undergraduates that may be expanded to include graduate students.

As the away-from-laboratory-work time that is required to perform a full internship can pose a problem for trainees (even if it is only 10—15 hours per week) and as the required time limits the number of such endeavors a single trainee can experience, we developed a parallel type of activity that we refer to as "site visits." These are experiential learning opportunities that consist of one- to four-hour on-site visits to an organization. The sites we have visited include biotechnology companies, laboratory incubator spaces, offices of scientific journals, a life sciences consulting firm, a technology and life sciences investment firm, and a community health center.

The goal of the internships and the site visits is to learn about career possibilities directly from employees in relevant work environments. They provide excellent opportunities for trainees to see a company's or organization's culture (vision, values, practice), the types of activities performed in different jobs, the variety of employment opportunities at a given site, and what each organization looks for when seeking job candidates. Internships and site visits can also be an opportunity to expand trainees' professional networks. These experiential learning activities further build on trainees' career exploration and preparation efforts.

Evaluation

In addition to our participation in the BEST awardees' national data collection project, local evaluation is focused on a variety of mixed methods. One strategy is to monitor web traffic to provide a count of total views of the Labor Insight™ data reports over time. Although imprecise, it can help to identify effective communication strategies that drive attention to the data.

We confirm the skill requirements associated with various job sectors by surveying the community through colleagues and alumni to determine if they agree with the critical skills identified using Labor Insight™.

We implement surveys to evaluate credit-bearing courses and accommodate based on student feedback if needed. Likewise, participation in workshops (number of attendees) impacts future offerings.

Assessments also include focus groups. In combination with interviews and surveys, focus groups have enabled us to learn about the many advantages associated with internships, such as how they give participants a realistic view of a job/career, of their performance, and of their potential for a future in the selected career path. Most trainees who participate in internships express an interest in continuing on the career path they tried out, but a small percentage realize that they are no longer interested in that specific path. In either case, trainees consider the experience a success as they sort through their career options. Focus groups have also demonstrated to us that site visits are very popular and that trainees generally consider them a good use of their time and effort, so we are likely to expand the site visit program in the future.

Conclusion

In summary, BU's BEST takes an evidence-based approach to determining workforce needs and to training graduate students and postdoctoral scientists to fulfill these needs as they seek satisfying careers in the biomedical sciences. Examining the workforce data contributes to a thoughtful development of coursework, panel discussions, and workshops, and to the identification of new site visits and internship opportunities. Trainees are encouraged to utilize the workforce data to make well-informed career decisions. As they consider their future goals, they are urged to meet with mentors to plan how to proceed by identifying needed skills and participating in BU's BEST programming to fill those skill gaps, ultimately enabling them to become more competitive job applicants. By the same token, the workforce is then supplied with applicants who are trained to meet employer needs.

An important component of the BEST initiative is to develop "best practices" for career advising and to share these with other institutions. We have developed a workshop that we present at conferences, such as the AAAS Annual Meeting and the Leadership Alliance SYNERGI Career Development Workshop, that we use to show others how the workforce data can be useful to trainees, faculty, staff, and administration. We have also shared the value of this approach with others at BU, which has resulted in the distribution of Labor Insight™ licensing costs with other groups within the university.

At every step along the way, evaluation helps change the framework of programming offered at BU, which allows for growth and development of BU's BEST offerings.

Acknowledgments

We would like to acknowledge our funding source, the National Institutes of Health, grant number 5DP7OD020322. We would like to specifically thank the BU's BEST Steering Committe including Lauren Celano, M.B.A., Co-founder and CEO, Propel Careers, Boston, MA and Kimberly A. McCall, Ph.D., Professor of Biology, Boston University, Boston, MA as well as Deborah M. Fournier, Ph.D., M.S., Assistant Provost for Institutional Research and Evaluation, Boston University School of Medicine, Boston, MA for her evaluation and assessment expertise. We would also like to acknowledge the BU's BEST Advisory Board. Finally, we thank Boston University for their institutional support of our program.

References

[1] https://www.burning-glass.com/products/labor-insight/
[2] https://www.bu.edu/best/job-search/biomedical-workforce-data/
[3] https://www.cirtl.net/
[4] http://myidp.sciencecareers.org Hobin JA, Fuhrmann CN, Lindstaedt B, Clifford PS. You need a game plan. Science September 7, 2012.
[5] https://www.bu.edu/arrows/
[6] https://www.bostontoastmasters.org/
[7] https://www.bu.edu/ctl/
[8] https://www.massbioed.org/events
[9] https://www.cirtl.net/events
[10] http://www.leadershipchallenge.com/professionals-section-facilitator-training-opportunities.aspx
[11] Kouzes J, Posner B. The Leadership Challenge. 4th ed. San Francisco: Jossey-Bass A Wiley Company; 2007.

Cornell BEST: Keys to successful institutionalization of career and professional development programming

Susi Varvayanis, Chris B. Schaffer, Avery August

Cornell BEST Program-Careers Beyond Academia, Cornell University, Ithaca, NY,
United States

On the shoulders of giants

Similarly to other institutions of higher education, Cornell University's Ithaca, NY, campus (referred to simply as "campus" throughout unless otherwise specified) is located in a rural environment that lacks obvious geographic opportunities for close interactions with major employers. In addition, its medical school, the Weill Cornell Medical College, is in New York City, roughly a four-hour drive away. Perhaps as a result of this relative isolation, Cornell has a rich history of easily forged intra-campus interactions and a truly collaborative environment. Pockets of excellent ideas, operating in a decentralized fashion, tend to be independently grown and curated across a broader life sciences environment distant from the New York City medical campus.

By the time the NIH solicited applications for the BEST Program, the aforementioned history had produced several successful models that brought broad disciplinary collaborators together. Offshoots of those programs taught us a few lessons that would then go on to shape Cornell's BEST proposal.

One model, the **Cornell Center for Advanced Technology (CAT)**, was created in 1983 and brings life sciences researchers together to work with industry and commercialize early inventions. Its long history of encouraging collaboration has spurred technology-based research and economic development in the state of NY. Part of a 15-institution consortium, the CAT has an educational component designed to encourage advanced

BEST: Implementing Career Development Activities for Biomedical Research Trainees
https://doi.org/10.1016/B978-0-12-820759-8.00002-4

undergraduates and first-year MBA students to do internships at NY-based companies and to foster science communication, technology transfer, and outreach; since the BEST award, eligibility has been expanded to include graduate students. Partnerships with large, state-wide companies were less feasible due to geographic distances. Instead of only measuring job impact, the CAT capitalized on small and startup venture creation for knowledge transfer. The **important lesson learned was to focus on Cornell's strengths** instead of trying to compete directly with other institutions. By tying our efforts to existing assets—namely, Cornell's educational excellence and interdisciplinary collaboration in biotechnology-related fields—a broader impact could be quantified.

The **New Life Sciences Initiative** was started in 2002 to integrate campus-wide life sciences with physical, engineering, and computational sciences, and involved 60 different departments and an unprecedented number of faculty hires. Through this new initiative, Cornell launched the Presidential Life Science Fellowship program, a fellowship for extraordinarily talented Ph.D. students, the following year with the goals of developing leaders for the future and exposing them to the variety of careers and skills they would need for their future success. Such skills include learning from professors about their career trajectories and mentoring philosophies, how they establish collaborations, and what they do outside the lab. **The lessons learned** centered on spurring new connections, facilitating meetings between graduate students and faculty with a broad disciplinary reach, and encouraging students to explore outside their field of entry across the life, physical, and engineering sciences **to broaden their horizons**. Encouraging faculty to **speak transparently** about mentoring and about their careers to help students determine which labs to select helped to ensure the best fit regarding mentoring approaches and goals.

Another successful initiative on which our BEST approach was based is **Entrepreneurship at Cornell**, which unites entrepreneurial endeavors across the Ithaca campus. Over the years, this virtual center changed the campus culture toward acceptance of entrepreneurship with some valuable lessons for any institution embarking on a similar mission. **The main lessons learned were to implement ideas with the most receptive groups first**, thereby demonstrating early successes to other departments/fields and fostering administrative buy-in, **and to focus on the customer**—in our case, the graduate students and postdocs—as this allowed us to assess demand, needs, and costs. In parallel, **this initiative underscored the importance of listening and adjusting actions** in response to faculty and administrative perceptions, bottlenecks, and frustrations.

The **Atkinson Center for a Sustainable Future** was established in 2007 as a result of a collaboration between faculty and outside partners. The group tackles the world's most pressing problems and offers us an additional lesson: faculty buy-in from across disciplines helps to create a more collaborative environment for research. The successful Atkinson Center model incentivizes tackling pressing societal issues with seed funding across disciplines with minimal administrative burden on the applicants.

At Cornell, graduate studies do not operate out of specific departments, but are instead organized into a field structure united by research interests spanning college and departmental administrative units. Graduate faculty from multiple disciplines and departments can associate with one or multiple fields so that students have access to broad, diverse scholarship and research in their respective areas of study. As a result, a particular biological science lab may have students from different departments, such as microbiology, immunology, nutrition, or biomedical engineering, engaging in research side-by-side. Given the breadth of the graduate field system and the lessons we learned from the above-mentioned centers on aligning many players to create a collective impact, it was only natural that our BEST proposal springboard from these transferrable lessons.

Scope

Stemming from a biomedical workforce report [1], the NIH BEST initiative focused on addressing broader training for biomedical doctoral students and postdocs. The NIH BEST solicitation announcement prompted Cornell to bring "allies" together to assess our existing capabilities and brainstorm programming ideas that would address apparent gaps. During this process, we recruited many faculty and staff from a variety of biomedical sciences and related disciplines who invoked their networks to garner a huge number of support letters from inside and outside Cornell. The tenets of the proposal focused on exposure to the breadth of career options where Ph.D.s are needed and on the opportunity to acquire skills and experience.

Although the effort at Cornell was initiated in the College of Veterinary Medicine, we leveraged the grant application process to get institutional buy-in from multiple colleges and the graduate school to enhance our chances of success. Since internal partners were fully supportive and deeply committed, this became an easy ask. Following the funding award announcement by the NIH, we amplified this pledged commitment by approaching the deans of the colleges that housed scientists and engineers on campus in Agriculture and Life Sciences, Arts and Sciences, Engineering, Human Ecology, and the Graduate School to get additional monetary buy-in. Since the faculty in these colleges had already benefitted from previous cross-campus collaborative efforts in the model programs mentioned above and since the graduate field system incorporates multiple departments, they were primed and eager to contribute funding to extend the BEST programs to all STEM fields for broader benefit.

A unique aspect of the Cornell proposal took advantage of the emphasis the NIH placed on the requirement that the BEST programs be "experiments" and proposed to retain a third of our operational funds for trainee-initiated ideas. These funds served to create maximal flexibility for taking advantage of future opportunities, for capitalizing on new ideas, and for entrepreneurially improving the program mid-course. Guidance for the funding requests is part of the experiential learning process for graduate students and postdocs, as it incorporates skill development for creating a budget, pitching an idea, and looking at competing offerings to ensure that their "ask" is outcomes focused and reasonable.

A small, dedicated group of faculty and staff that had a vested interest in a successful proposal created a cohesive and positive message that was dedicated to the long-term goal that Cornell BEST become institutionalized. The proposal focused on culture change and on garnering widespread acceptance of the program to help accelerate its early growth.

These strengths were recognized by reviewers, who indicated that,

> The nature of the graduate training program at Cornell University is inherently very flexible, potentially multidisciplinary and has a collaborative environment that lends itself to this proposal to take advantage of this environment. There is a plan to also take advantage of environments outside the training institution, which will enhance the educational value of this program.

They further recognized that the program would provide trainees with the opportunity to learn about and obtain practical experience in every career area requiring scientific expertise, a model that could also be implemented in other institutions regardless of size or location. In addition, our concept for hands-on team projects was, in the words of the reviewers, "unique."

The numerous letters of support from on- and off-campus partners evidenced our enthusiasm for creative ideas and broad collaboration. As an example, Entrepreneurship at Cornell provided Cornell BEST broad access to its extensive list of mentors drawn from prominent alumni from the SC Johnson Graduate School of Management. These potential mentors are available to any Ph.D. student or postdoc in our program who is interested in business or consulting—an unprecedentedly open attitude in the business school sphere.

Reaching our audience

To realize full institutionalization of professional development and experiential programming in the long term, it was important to get broad participation. Key aspects to attain this goal included holding open events, advertising widely, and collaborating with on-campus departments, student- or postdoc-led organizations, and off-campus employers and trade organizations to address gaps in programming. We advertised using existing mechanisms for communicating events to doctoral students and postdocs, particularly through Graduate Field Assistants, staff members who assist the Directors of Graduate Studies, are familiar with campus resources, and are generally considered the first point of contact for student questions. We also partnered with the Office of Postdoctoral Studies to reach postdoctoral researchers via their regular notices distributed by a listserv. The full communication plan included advertising the kickoff event by sharing a simultaneous press release about the program in the university news [2] and a link to an interview with faculty and key personnel in a scientific journal [3]. The development of a dynamic website [4] fostered wide participation by providing public access to resources.

Strategic follow up with participants included sending them a link to a newly developed application/survey to invite their involvement in the BEST Program and subsequently holding individual consultations with them to determine their needs and evaluate potential Advisory Board recruits. The resultant BEST Advisory Board, made up exclusively of graduate students and postdocs, went on to shape the program and provide needs assessment and boots-on-the-ground feedback for events. Furthermore, their involvement would provide BEST with the hands needed to assist with programming while they explored careers and gained skills needed for their future.

Defined approaches with built-in flexibility: Developing Cornell BEST

We initially made plans to **address what we thought were skill gaps** based on existing survey literature and conversations with employers within MedTech, New York's trade organization for the bio/med industry, of which Cornell University is a founding member [5]. As co-chair of the MedTech Workforce Development Committee, BEST Program Executive Director Susi Varvayanis was able to collect survey data from among the 100 MedTech members to better understand what they needed from their employees. Three themes emerged as to which skills needed improvement: effective communication across all levels, teamwork, and knowledge of industry practices.

Communication

Each year since 2007, Cornell Professor of Science and Technology Studies Bruce Lewenstein teaches a weekend workshop on communication. Thanks to the NIH BEST grant funding, we immediately increased its frequency to twice a year. Originally developed as a full-semester course, it was distilled down to a one-credit, intensive weekend workshop due to the expectations that grad students' and postdocs' advisors had on how much time their trainees should spend outside of the lab. This change made the program more attractive and increased the number of faculty that referred the workshop to their own students and postdocs.

In addition, we developed a workshop called "Finding your Research Voice." Taught by Cornell Professor of Physics Itai Cohen and the Chair of the Department of Drama at the University of Alberta, Melanie Dreyer-Lude, the course uses improv techniques and storytelling practices (such as the dramatic arc) to help participants develop better analogies and visuals for the first 10 minutes of their own research talks. As part of the workshop, participants receive peer and video feedback to improve their voice and body language to better engage their audience.

Teamwork across disciplines

Immediately after receiving funding, we leveraged the successful Pre-Seed Workshop [6] and convinced the organizers to allow a BEST participant to join each workshop team to address the need for effective teamwork across disciplines. The Pre-Seed Workshop is a fast-paced, team-based, hands-on workshop that brings professionals together to guide a university researcher who has a high-tech idea they think might be worth commercializing. Each team consists of a hand-picked intellectual property lawyer, a financial expert, a business professional, a seasoned entrepreneurial coach with a background in the technology to be evaluated, and an M.B.A. student or BEST trainee. The teams hold 9 analysis sessions in which they address 20 key questions to determine whether a particular high-tech idea is worth commercializing. We found that the richer the mix of backgrounds (across the life, physical, or social sciences), the stronger the end products and commercialization plans were. Since then, we have incorporated the inclusion of participants from multiple fields and disciplines as a best practice for almost every experiential activity we support.

To develop a culture more focused on teamwork across disciplines, we initiated a collaboration with a startup company to customize market research and to include Ph.D.'s and postdocs on their cross-disciplinary teams undertaking projects for real companies. The feedback from this collaboration and a review of the cost-to-benefit ratio indicated that although it was useful for some individual projects, most participants indicated that it was not specific enough for Ph.D.-level participants. As a result, the program was terminated.

Prior industry experience

Biotechnology companies prefer Ph.D. hires who have prior industry experience, which trainees often lack. A focus group with employers and trade organization members identified several gaps in trainees' knowledge, including a lack of training in regulatory affairs (this training was only available on an as-needed basis in formal Master of Engineering design projects). Thus, our course-planning efforts focused on introducing trainees to topics such as quality control, quality assurance, and FDA regulations in the context of the biopharma industry through the use of specific course materials and a lineup of guest speakers and lecturers.

Listen to your customer

To provide the most productive guidance to participants, we developed an **application to determine their interests** [7]. The goal was to capture participants' demographic information, interest in various career tracks (e.g., communication; policy; governance, risk, and compliance; and industry, entrepreneurship, and management), and hopes for what they wanted to get from their involvement. The latter information was helpful for early marketing to other applicants and to faculty, and it further helped guide event planning to personalize the program to common needs.

The data from this application process showed that contrary to our extensive plans to develop significant offerings in regulatory affairs, only 1.7% of the initial applicants were interested in this area. We also found that 46% of the first group of applicants were very interested in consulting, an area that we had not initially considered. The questionnaire also included questions about the use and utility of existing services (e.g., consultations at the Office of Postdoctoral Studies, Career Services Office seminars, the Postdoctoral Leadership Training workshop, etc.) so that we could provide additional feedback about successful activities to our colleagues, leading to potential co-sponsorships or discussions on how to improve/address other areas that were not meeting existing needs.

Through this early process we learned a valuable lesson applicable to any institution endeavoring to develop a new program in experiential professional development. **Let the graduate students and postdocs tell you what they need** early to avoid investing too much time and effort in areas that will not be of value to them. Thanks to this realization, we learned that a very small percentage of the initial applicants were interested in the area of governance, risk, and compliance, so we scrapped the idea for a course in this area and pivoted to find alternative options.

We discovered that TechnologyEd [8], a member of the American Association for Adult and Continuing Education (a consortium of academic and corporate partners that offers on-line instruction in various industry-related compliance areas), hosts many online courses designed by industry experts. Thanks to the fact that we had set aside some funds to accom-modate shifting needs, we supported a few participants who wanted to take some of the available courses, which include topics like biopharmaceutical regulatory affairs, good clin-ical practices, and quality assurance and quality control. The trainee reflections and feedback revealed the value of these courses to us, and we subsequently supported other trainees who also wanted to participate. By listening to the needs of our trainees, we avoided creating a course that would likely not have been well attended, and instead made use of an existing resource by allowing trainees to participate and reimbursing them upon course completion.

Lessons learned

The main lesson we derived from our experiences was to **pivot early to embrace the mar-ket need and to modify our programming plans after receiving feedback**. This flexible pro-gramming was designed to cater to the changing market need (of both employers and trainees), which was essential for our small program to sustain a successful model while requiring the smallest possible investment of time by our staff.

In the spirit of experimentation, responsiveness, and flexibility, the growth of the Cornell BEST Program was driven by these principles, allowing us to **find our niche to fill gaps** in previously implemented programming. Through the participant application and surveys, in-dividual consultations, and discussions with faculty and colleagues, we gained an under-standing of where our participants found value. However, some themes in our program were built on the experience of faculty participants. For example, Chris Schaffer, a PI on the BEST grant and Associate Professor of Biomedical Engineering, served as a AAAS Science and Technology Policy Fellow during his sabbatical leave and as a result came back to Cor-nell with a new respect for the value of excellent science communication and for the impor-tance of science input in policymaking. Upon his return, he developed the "Science Policy, from Concept to Conclusion" course with early help from a faculty grant awarded by the Engaged Cornell program. Schaffer and a postdoc from his group, Catherine Young, devel-oped the course to help address the lack of policymakers who have significant scientific or engineering knowledge and the need for scientists who have experience with the inner work-ings of public policymaking. The course, taught through a project-based learning approach, brings together teams of STEM students to develop expertise in an issue of their choice that sits at the intersection of science and public policy, develop a policy prescription, and "go live" by advocating a policy change to local, state, or federal policymakers.

In part because of the enthusiasm of students who had taken this policy course, the BEST Program supported the establishment of Advancing Science and Policy (ASAP), a graduate student-run organization. This group is dedicated to training STEM researchers at all career stages to share the value and significance of their work with policymakers. Members develop skills and professional networks by attending regular workshops, faculty chats, and talks by invited speakers. Through these resources, they learn to communicate with policymakers and the general public about the role of scientific input in advancing policy and society. The organization serves as a hub to bring like-minded participants together who take ownership of the programming and develop new experiential learning opportunities for others to benefit. The "Take a Politician to Work Day" effort implemented by graduate student Leah Pagnozzi [9] is an example of an outcome that resulted from a BEST participant taking the science policy course and being an active member of ASAP. Other ASAP efforts include supporting writing initiatives and leading trips to Albany, NY, and Washington, D.C., to meet with policymakers and learn firsthand about policymaking and advocacy. These experiences set the stage for successful applications to fellowships such as the AAAS Science and Technology Policy Fellowship.

Responsiveness to participant and faculty interests

In response to the high number of participants from the early application pool who expressed an interest in consulting (46%), we focused our support on a student-led graduate organization called the Cornell Graduate Consulting Club (CGCC). This group is distinct from the MBA-led consulting group within the business school, which does not address the unique needs of Ph.D.'s well. CGCC brings in speakers who have transitioned from research into technical or management consulting careers, individuals in positions to hire Ph.D. consultants, recruiters who are willing to look over application materials, and alumni in consulting who can help trainees practice case interviews.

Another gap was identified as a result of early feedback sessions with business school faculty who had had Ph.D. students or postdocs in their MBA classes. These discussions revealed the need for us to prime our students with the vocabulary of the business world to help them understand the relevant jargon and avoid basic questions that tended to slow down the class since the MBA students were already familiar with such terms. We developed a mini-course called "Business as a Second Language: From Molecules to Moolah" to help Ph.D.'s and postdocs develop this business and management vocabulary through the use of cases relevant to their fields and pitch presentations as part of the final exam. This offering evolved into a full course taught by Robert Karpman, who is now Professor of Practice and Director of the Minors Program for Life Sciences (a business minor for life sciences majors) in the Dyson School for Applied Economics and Management, part of the SC Johnson College of Business. The course, "Business and Management Fundamentals [10]," teaches basic business and management skills for career advancement. More recently, Karpman created a seven-course online certificate program in business excellence for healthcare professionals [11] as part of eCornell, Cornell's online learning platform. Geared toward STEM disciplines, course outcomes include determining the value of a new biomedical technology from a business perspective, preparing and presenting a strong business case for a new device or technology, and explaining the difficulties in getting new technology into the healthcare marketplace.

Another best practice we discovered early on was to **collect (enough) data to be convincing to the program organizers, participants, and faculty**, which allows for nimble responsiveness. After the first and second years of BEST, we comprehensively surveyed

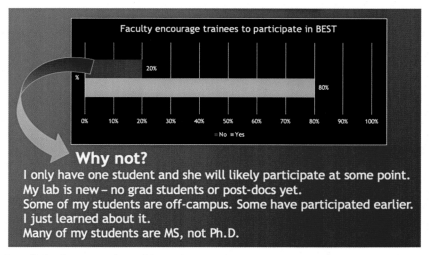

FIG. 2.1 Cornell faculty respondents (N = 128) voice their encouragement for trainee participation in BEST activities.

the participants on the activities they considered to be worthwhile for the time involved in doing them and on the skills they learned that they felt were the most important for their career success. The data revealed which activities should be of higher priority to justify our investment in them and which activities were low-value programming that we ought to discontinue due to little perceived value.

One and a half years into the program, we surveyed faculty to gauge awareness of BEST programming and to determine whether they would refer their advisees to participate (Fig. 2.1). We also asked them what activities (Fig. 2.2) and skills (Fig. 2.3) they thought would benefit our trainees the most, allowing us to reorganize our priorities where appropriate.

We found that faculty considered communication skills important (Figs. 2.2 and 2.3); therefore, we continued to focus on communication by providing the weekend work-shops twice a year and developing the "Finding Your Research Voice" workshop (both described above). We also determined from the faculty survey that outreach and connections with recent and more experienced alumni were key. Faculty priorities matched what participants told us were the most useful skills in separate annual sur-veys. This finding reinforced our confidence in directly approaching the "customers" to determine their needs.

Conducting faculty surveys like this one also helped to increase awareness of BEST. To avoid survey fatigue, we distributed the surveys indirectly and informally via department chairs, who then shared them with their faculty who advised grad students and/or postdocs. We also asked faculty who had written support letters for our NIH proposal to share the sur-vey with their colleagues. Notably, no reminders were sent. Using this approach, we received more survey responses from faculty (128) across many disciplines than the initial number of invitations (100).

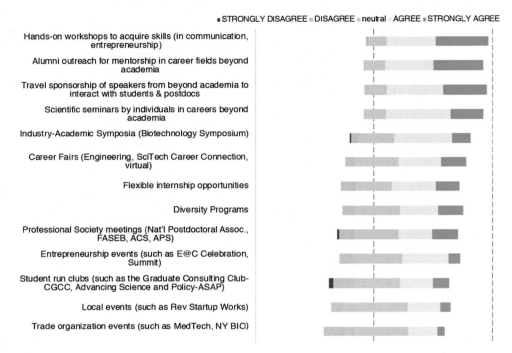

FIG. 2.2   Cornell faculty respondents (N = 128) voice their level of agreement with the types of activities they think BEST should focus on.

A focus on enhancing experiences

Courses are not the only method of addressing gaps in knowledge of the workplace. We developed several short-term, immersive site visits ("treks") based on the fact that graduate students and postdocs in the BEST Program long to see and experience the actual work environment of Ph.D.'s in the field, that faculty expressly desire to not have their students away from their research any longer than necessary to make good career decisions, and that the NIH has warned against lengthening the time to degree through BEST Program activities. These treks are designed in partnership with industry, led by faculty, or run entirely by students with guidance from BEST staff.

To give an example, the Silicon Valley Trek is led by Michael Roach, the J. Thomas and Nancy W. Clark Assistant Professor of Entrepreneurship at Cornell, who was an incoming professor at the time. This trek was a collaboration between BEST staff and Roach, whose research on Ph.D. career preferences and decision making overlapped with our objective of enhancing the self-efficacy of graduate students [12]. Cornell BEST and the College of Engineering each sponsored four trainees who wanted to participate in the Silicon Valley Trek. Capitalizing on Roach's existing network, the trek added value to the participants' experiences by introducing them to a number of senior colleagues from broader disciplinary backgrounds. The four-day trek included visits to several accelerators, meetings with

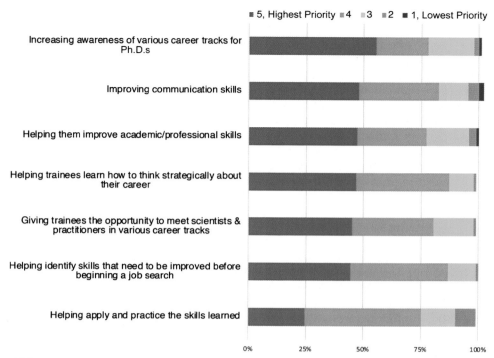

FIG. 2.3 Cornell faculty respondents (N = 128) share their priorities for BEST skill development.

entrepreneurs (most of whom were alumni who had started businesses as graduate students), and attendance at a coincident alumni celebration event.

By speaking with founders during the treks, participants gain firsthand insights into the benefits of the particular region, receive advice for their future endeavors, and see how and where the work is accomplished. One participant indicated that the trek they were on was by far the most impactful thing they had done in their five years at Cornell, as it galvanized their desire to become an entrepreneur but helped them realize that they would not want to do so in Silicon Valley. This individual went on to start their own company based on their dissertation research near a national lab in Tennessee. We realized early on that a 'rule-out' decision outcome like this one is as essential as realizing one's passion, and that these short site visits can provide sufficient insight to support either conclusion.

Another sample trek is the Drug Discovery Conference, which was co-designed by Avery August, a BEST PI and Professor of Microbiology and Immunology, and Cherié Butts, Associate Medical Director in Clinical Development at Biogen [13]. This conference stemmed from Butts's prior participation in a Cornell BEST symposium and from a site visit coordinated by a BEST Program participant that involved graduate students shadowing partners in various units of the company. The three-day Drug Discovery Conference was designed to serve multiple institutions and expose participants to a variety of roles that support the drug development pipeline. Discussions and hands-on experiences help to demystify the new drug and biologics approval process and teach how products of interest are evaluated for licensing.

Another very effective method for participants to experience the work that someone with a STEM Ph.D. can do outside of academia is the "BESTernship." These experiential opportunities consist of embedding individuals in campus-based offices to engage in part-time experiential opportunities (5–10 hours/week) in areas such as science communication, business and entrepreneurship (through the McGovern Center for Venture Development in the Life Sciences), or regulatory affairs (through the Environmental Health and Safety Office), to name a few. Thanks to their BESTernship experience with mentor Alexis Brubaker, the Biosafety Officer of Cornell University, two Cornell students applied for and become finalists for the prestigious National Biosafety and Biocontainment Training Program (NBBTP) fellowships, which are awarded to one to four individuals internationally each year. At the time of writing, one of the students had been awarded the fellowship.

Cornell University's BEST program attracts and engages participants because of the content of the programming, not because of the perks. We advertise the program to Ph.D. students and postdocs as a valuable resource that aligns with their goals of learning about career-related opportunities and gaining experience for their resumes, not as a source of free lunches or purveyor of financial resources. Accountability is another important aspect of our relationship with participants; once individuals have signed up for an opportunity, we expect them to follow through with participation or to send a note indicating their anticipated absence. Clearly communicating our expectations for our participants' professionalism further vets them for future employers.

Transition from external to internal funding model

Three and a half years into the program, we embarked on a cross-campus departmental roadshow to present to faculty our evidence of broad interest in BEST programming, share the challenges we had encountered thus far, and seek input on how to sustain the programming. This process served several purposes: it increased awareness of what the program had accomplished, helped us gain strategic input from faculty on carving a path forward, and increased broad awareness and buy-in for the continuation of funding. As a result of these activities, we were encouraged to increase the visibility of our higher-level successes via spotlights on the BEST website [14], our LinkedIn group [15], and Twitter [16], and through the creation of a newsletter where we could broadly share our outcomes with faculty. We published several articles in the university's official news outlet, *The Cornell Chronicle*, and invited existing BEST participants to speak at student orientation events.

The ongoing collaboration with the Cornell Graduate School to harmonize all professional development programming for all graduate students and postdocs promised to ease the transition from being supported by the NIH grant and the colleges to receiving new, sustained funding from the provost. This new funding will actually broaden the scope of the program to serve not just STEM trainees but to include the social sciences, humanities, and arts going forward.

Collaboration with other academic institutions

A very important aspect of being part of the NIH BEST Consortium was the free sharing of resources across the institutional constellation. For example, at the start, Cornell BEST adapted an application developed by New York University's BEST Program, who freely shared their initial questionnaire with us for our own use. Through it, we obtain basic demographics, career interests, motivation for participation, and perceived advisor support of participants interested in Cornell BEST. Although we made some modifications, keeping many of the same questions facilitates cross-institutional comparisons between us and NYU, allowing us to evaluate the broader impact of our findings.

The costs and time saved by sharing of our successes and challenges at the annual in-person meetings and the monthly conference calls with other consortium members are irreplaceable. Many examples can be seen on the NIH BEST website. We learned that "stealing" ideas from and sharing approaches, results, and failures with institutions within and outside the NIH BEST Consortium saved us time when developing programming tools, especially when the institutions we worked with were similarly committed to a culture change—that is, to exposing trainees to careers outside of academia and increasing the institutional acceptance of these career paths.

Challenges

As with any startup, one of the biggest challenges is scaling up, an inherent problem of a successful endeavor. This was addressed to some extent by leveraging collaborations and resources through partnerships such as those with the SC Johnson College of Business and other on-campus offices. Another example of how we have scaled up involves career consultations with trainees, which were previously held on an individual basis and have now morphed into small-group consultations. These small groups improve the sense of community in participating trainees, increase cross-disciplinary interactions, and save staff time.

As we move forward, we are exploring ideas for better interdisciplinary integration of the humanities and social sciences with the life sciences, physical sciences, and engineering. One early success has been the aforementioned and trainee-run CGCC, which is now co-led by a graduate student in Food Science and Technology and a graduate student in Near Eastern Studies who share a passion for the same career track. Other successes include developing programming that is inclusive of all fields. As a practice, we always encourage partners and co-organizers of activities to broaden to aligned fields and to co-sponsor the participation of individuals who normally fall outside their departmental norms. For instance, trainees from Integrative Plant Science or Applied and Engineering Physics might be interested in attending a career panel discussion on industry careers hosted by Microbiology, and we help facilitate their participation.

As we start thinking of how to become "bigger and better," we will have to balance our efforts of providing customized programming and experiential opportunities for individual fields with areas where we can bring more fields together without diluting the relevance. We have started by determining where natural alignments lie: Linguistics with Computer

Science and Philosophy; English with Comparative Literature, Romance Studies, Near Eastern Studies, and History; Statistics with Computer Science, Mathematics, Applied Mathematics, Information Science, and some English fields; and so on. These graduate field alignments will foster future efforts to collaborate across student-run organizations and will provide broader programming and advertising channels. Perhaps thinking about creating a "universal design" will help other institutions frame their career and professional development programming to be inclusive from the start.

Going forward, we will continue to refine our approach and to learn from our failures so that we can provide experiential learning to the graduate students and postdocs at Cornell and empower them to create their own opportunities for their future career successes.

Acknowledgments

Cornell BEST was funded by the National Institutes of Health: Strengthening the Biomedical Research Workforce award number DP7OD18425. Institutional funds were provided by Cornell's Provost to BEST to help with the program's university-wide expansion; "Careers Beyond Academia" will be the new name for Cornell's BEST Program in the Graduate School starting in the Fall of 2019. Additional thanks to the Elsevier editorial team and to Roger Chalkley.

References

[1] National Institutes of Health. Biomedical research workforce working group report. 2012. Available from: https://acd.od.nih.gov/documents/reports/Biomedical_research_wgreport.pdf.

[2] Ramanujan K. BEST program will train Ph.D.s for nonacademic careers. Cornell Chronicle; 2014. Available at: http://news.cornell.edu/stories/2014/03/best-program-will-train-phds-nonacademic-careers.

[3] Shtessel L. NIH program trains scientists for nontraditional careers. Science Careers 2014. https://doi.org/10.1126/science.caredit.a1400050. Available from: https://www.sciencemag.org/careers/2014/02/nih-program-trains-scientists-nontraditional-careers.

[4] Cornell BEST website: http://best.cornell.edu/. This site will be migrating to gradcareers.cornell.edu as it expands to a university-wide level.

[5] MedTech website. Available from: https://medtech.org.

[6] Pre-seed workshop website. Available from: http://www.preseedworkshop.com.

[7] Cornell BEST Application. Available from: https://cornell.qualtrics.com/jfe/form/SV_b1QtNWOWgrIfGx7.

[8] TechnologyEd website. Available from: www.technologyed.com/online-courses.

[9] Jennings, C. Featured force: Leah Pagnozzi. Available from: www.aaas.org/featured-force-leah-pagnozzi.

[10] Cornell University Registrar Courses of Study, Office of the University Registrar. AEM 6145- Business and Management Fundamentals for STEM Graduate Students. Available from: http://courses.cornell.edu/preview_course_nopop.php?catoid=33&coid=530518.

[11] eCornell Certificate Program, Business Excellence for Health Professionals. Available from: www.ecornell.com/certificates/healthcare/business-excellence-for-health-professionals/.

[12] Saurmann H, Roach M. Science PhD career preferences: Levels, changes, and advisor encouragement. PLoS One 2012;7(5):e36307. Available from: https://journals.plos.org/plosone/article?id=10.1371/journal.pone.0036307.

[13] Butts C, August A. A novel target for accelerating drug development: Biomedical science training. 2018. Available from: www.sciencemag.org/advertorials/novel-target-accelerating-drug-development-biomedical-science-training.

[14] Cornell BEST website. Available from: http://best.cornell.edu/.

[15] Cornell BEST LinkedIn Group. Available from: www.linkedin.com/groups/6655233/.

[16] Cornell BEST Twitter feed. Available from: https://twitter.com/Cornell_BEST. *[Please follow us!]* .

The Atlanta BEST program: A partnership to enhance professional development and career planning across two dissimilar institutions

Tamara Dahl[a], Rebekah St. Clair[b], Julia Melkers[b], Wendy C. Newstetter[c], Nael A. McCarty[a, d]

[a]Laney Graduate School, Emory University, Atlanta, GA, United States; [b]School of Public Policy, Georgia Institute of Technology, Atlanta, GA, United States; [c]College of Engineering, Georgia Institute of Technology, Atlanta, GA, United States; [d]School of Medicine, Emory University, Atlanta, GA, United States

Introduction and approach to the problem

The primary focus of biomedical Ph.D. training is the generation of new knowledge to advance science, medicine, and health. Primary skills developed in Ph.D. training are critical thinking and the ability to make hypothesis-driven inquiries that lead to sound experiments and reliable conclusions. Given the way science is done today, with researchers combining a variety of methodologies in projects managed as a team, the skills gained in Ph.D. training can be leveraged in a variety of work settings outside of academic laboratories. Critical thinking and the ability to make informed recommendations are essential in not just the execution of scientific research, but also in sectors of the economy where science is regulated, commercialized, practiced, and used by everyday people.

Scientists do not lose their identities or the skills they developed as Ph.D. students or post-doctoral fellows when they take off their lab coats to enter "non-traditional" positions outside of the academic lab. The training they receive and the experiences they have, including

BEST: Implementing Career Development Activities for Biomedical Research Trainees
https://doi.org/10.1016/B978-0-12-820759-8.00003-6

organizing and completing a dissertation and engaging with diverse scientists from an array of academic and cultural backgrounds, will stay with them. Although they will always inherently be scientists, they may apply their skills in careers and sectors that they are well suited for outside of academic research given their personal strengths and well-informed career interests. For example, trainees might discover that they are better suited for negotiating licensing contracts to get drugs into patients rather than for coming up with the next novel advance in neurobiology. Both roles are critically important to making scientific progress and improving health outcomes.

In every sense, this broader perspective about what it means to be a scientist in today's modern biomedical workforce is a good thing. It's a good thing for the trainee because it enables them to plan for and attain long-term career satisfaction, and it's a good thing for the faculty because their trainees can identify and prepare for a career for which they feel best suited and can spend more energy being productive in the lab rather than dreading their next steps after graduation. Having a clear path forward after training helps with trainee motivation and can keep trainees committed to attaining their research goals.

This broader perspective about one's role in science is also important for the scientific and innovation enterprise at large. Preparing trainees to think broadly about their research across scientific and related disciplines (such as big data and business) stimulates the national economy. In addition, broadening skill sets to include self-awareness, leadership, and stress management lays the foundation for lifelong professional development, putting trainees on a path that prepares them to be confident, excited, and capable in their next steps after their Ph.D. training.

Current Ph.D. training is largely based on an apprentice model that prepares trainees to become tenure-track principal investigators (PIs) in academic research. At the same time, the number of new tenure-track faculty positions lags well behind the number of Ph.D. trainees that graduate every year. We also know that the rates of unemployment in this same population of young, biomedically trained scientists is very low, so our trainees are not struggling to find jobs; they are, however, struggling to find the jobs that are the right fit for them and to gain the skills needed to be competitive and competent in those jobs.

As science continues to change, so must training, education, and academic communities. Similarly, as funding agencies change their objectives and criteria for awarding training grants to include professional development activities, top research universities need to hire staff and provide quality resources dedicated to the professional development of modern scientists, who have a richer and more diverse set of career options available to them compared to traditional career options.

Ph.D. trainees must continue to receive field-specific research training, focusing on the central research fundamentals and core competencies of critical thinking, problem solving, technical skills appropriate to their specific field, communication for scientific audiences, and generation of new knowledge. However, Ph.D. training also should include training that touches on the following three foundational aspects:

- **Developing Self.** Becoming the best version of themselves, by gaining skills in:
  - Self-awareness and management of personality styles/needs
  - Individual development planning
  - Time/project/stress management

- **Developing as a Student.** Becoming a scientist, by understanding:
  - Interdisciplinary research
  - Academic environments
  - Research in non-academic sectors
  - Mentorship
- **Developing as a Professional.** Becoming a professional, by focusing on:
  - Leadership and management
  - Career exploration and development

Atlanta BEST, and our goals

The National Institutes of Health (NIH) launched the BEST initiative to accomplish these goals [1,2]. Emory University (Emory), in collaboration with the Georgia Institute of Technology (Georgia Tech), received a BEST award in 2013 and established the Atlanta BEST Program. Outcomes from the Atlanta BEST Program and lessons learned are described here. Although the structure of the BEST Program per se is not being incorporated into either institution, the new understanding gained through the experiences of the BEST Program, as outlined below, is continuing to impact graduate training at both institutions.

The goals of Atlanta BEST were to provide broader and more effective training by enabling positive student and postdoc participation and outcomes that impact how they value and reflect on their Emory/Georgia Tech training experience. Desired outcomes for trainees included being informed about career options that leverage Ph.D. training; participating in experiential career development opportunities; building skills around career planning and individual development; being confident, capable, and excited about next steps after training; being competitive and prepared for the jobs they want; and attaining long-term career satisfaction. Desired institutional outcomes included strengthening a reputation for training well-rounded and successful graduate students; developing networks of engaged alumni; and improving faculty and student recruitment, productivity, and retention.

Shifting the landscape of Ph.D. training to incorporate shared roles

Traditional Ph.D. training landscape: Faculty-centric training model

Training in research fundamentals is the core venue by which our students earn the Ph.D. degree. However, employers that seek the high-level competencies of a Ph.D. holder find that graduates often lack skills necessary for non-academic environments. Although Ph.D. scientists leave their doctoral training with the ability to work with complex problems, technical skills, research expertise, and the ability to apply new knowledge, leaders in STEM employment sectors have indicated that Ph.D. graduates lack competencies in areas such as: data science; science policy; technology commercialization; governance, risk, and compliance; time management; project management; working in teams; presentation skills; and non-academic communications.

The current, traditional model of graduate Ph.D. training is centered upon a faculty member — the PI — who mentors trainees with the expectation that they in turn will become

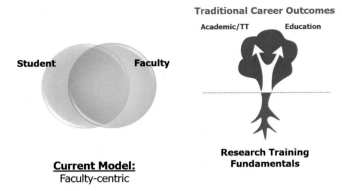

FIG. 3.1   Current model of Ph.D. training. Mentoring and training are provided primarily by the PI. Rooted in the classic research training fundamentals, this model is best designed to train students for faculty careers in academic research or education.

tenure-track (TT) PIs at an academic institution (Fig. 3.1). In this model, the primary career outcome is the continued engagement in research and education. The PI and the trainee participate in an apprenticeship in which the PI draws upon his/her own knowledge and experiences and is the primary source for trainee professional and scientific development. The in-depth, advanced training that trainees receive from PIs develops into the fundamental competencies and research skills that are the foundation of a Ph.D. program.

This faculty-centric model is limiting and uneven because of the variable skills, style, experiences, and interests that PIs have. Many only have experience in academic environments and thus do not have the knowledge or networks to guide trainees in broader career exploration or development of skill sets outside of their areas of expertise.

In addition, our Ph.D. programs, departments, and universities play a small role in trainee career and professional development in the current model. For those trainees pursuing traditional TT faculty positions, this model may be sufficient. However, trainees wanting to explore and pursue broader career interests have little relevant mentoring and resources available to them. As a result, career planning and professional development efforts are often put off until the last minute, and often are pursued with little guidance or confidence.

It is important to note that exposure to broad career and professional development training is beneficial in many respects regardless of an individual trainee's career interests. Making informed and strategic career decisions will benefit trainees whether they become traditional faculty members or work outside of the lab.

Future Ph.D. training landscape: Shared roles

As Fig. 3 1 shows us, there is not much room for larger institutional roles to formally fill in the gaps of broader career exploration and professional development. The development of research skills and perspectives are — and should remain — of the utmost importance for Ph.D. training. However, graduate training needs to incorporate broad-ranging, formal programming and resources toward enhanced career **self-efficacy**, a person's belief in their

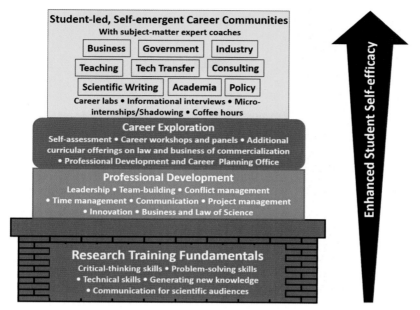

FIG. 3.2 The student training continuum: research training + professional development + career exploration and planning. As we build upon the foundation of research training fundamentals, students experience enhanced self-efficacy and agency toward making informed career decisions.

abilities to be successful. Centralized university resources can provide additional, broader training that matches desired career outcomes and workforce needs.

Students must continue to receive field-specific research training with a focus on the tried-and-true fundamentals of critical thinking, problem solving, technical skills, communication with scientific audiences, and generation of new knowledge. However, we must build upon that foundation by providing professional development training and opportunities for career exploration that will enhance our students' efficacy toward both making informed career decisions and successfully transitioning to those career outcomes (Fig. 3.2).

Most PIs are not able to (nor should they be expected to) prepare students for the broad array of possible careers available to their trainees, nor are they able to train them in diverse skill sets beyond their areas of expertise; therefore, other components of the training ecosystem must provide programming that fills in the gaps. Although there is a clear need for the trainees to take more ownership of their own training and development, an important role also exists for institutional stakeholders to provide non-laboratory/research training and career resources (Fig. 3.3).

At a high level, the shared-responsibility model would provide formal, institutional opportunities for:

- **Broad exploration** of self and of career options,
- **Enhanced development** of interpersonal and career-related skills, and
- **Opportunities and action** in planning and experiential learning

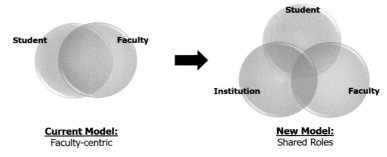

FIG. 3.3   Shift in Ph.D. training models. In the new model, the PI provides scientific training and development that is the foundation of the Ph.D. degree, but the students and the institution have roles in driving the students' development and in filling in the gaps with broader training and resources that complement research training and development, respectively.

As a result of implementing this shared-roles model for Ph.D. training, trainees will become able to identify additional training needs and to make informed decisions about their time as a trainee and their future career paths. In the meantime, the PIs will remain at the center of the training efforts, continuing to build and develop the trainees' core scientific skills. The requirements and expectations determined by faculty and departments or programs will not change or diminish when it comes to scientific rigor and productivity. The benefit of this shared model to PIs is that it shifts some of the training burden off of faculty while providing them with formal, tested programming in mentorship intended to improve their abilities to guide trainees with respect to executing their own training and career plans.

Incorporating a role for the institution in graduate student training begins to address some of the challenges that faculty and students face regarding professional and career development, such as adequately understanding the diversity of career opportunities available, offering resources to develop skills necessary to be successful in the trainee's career of choice, and building broader networks.

When it comes to professional development, academic departments and graduate programs may be better suited for building and maintaining field-specific alumni and employer relations; combining these resources with non-academic resources can help identify and develop programming and tools that keep up to date with current trends in various rapidly changing career sectors. Importantly, input from individual programs can help ensure that resources provided at the institutional level are relevant to their students. Faculty mentors and graduate programs can't do this without institutional resources and commitment. By taking a broader approach to the development of our graduate students and postdocs, which adds professional development activities and career development activities to these ongoing research training fundamentals, trainees can be exposed to a broader set of possible career outcomes and will be effectively prepared to enter them (Fig. 3.4).

The later sections of this chapter outline the interventions and lessons learned from the Atlanta BEST Program, from when the program officially launched in March of 2014 through July 2018.

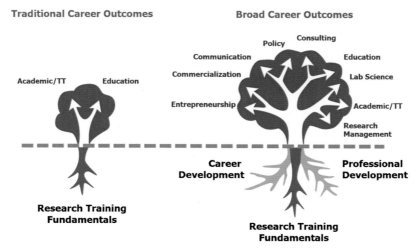

FIG. 3.4 Training that only focuses on research fundamentals only prepares our trainees for traditional career outcomes. By adding activities in professional development and career development and exploration, we make it possible for our students and postdocs to know about the wide variety of potential career outcomes and to be prepared to make informed decisions for their own career futures.

The Atlanta BEST program

With five-year funding from the NIH BEST initiative for the Atlanta BEST Program, Emory and Georgia Tech explored resources, programming, and approaches to identify interventions with the greatest impacts on Ph.D. trainees' ability and confidence to make informed career decisions. Some desired outcomes of our and other NIH-funded BEST programs included:

1. Improved trainee understanding of available career opportunities, confidence in making career decisions, and improved attitudes toward non-academic career opportunities.
2. Reduced time to achieving desired non-training, non-terminal career positions (vs. a postdoctoral fellowship, for example), and reduced time in those positions.
3. Creation or further development of institutional infrastructure to continue BEST-like activities after the funding period ended.

The Atlanta BEST Program included both trainee-focused efforts and faculty-focused efforts, as represented by the aims in the original NIH grant (Box 3.1). It was designed around a cohort model in which Ph.D. students and postdocs in the biomedical research space were engaged for two years. The first year was focused on exploration of a breadth of career directions and the second year was focused on independent immersion into one or more of the six "career tracks" shown in Fig. 3.5, along with additional training relevant to traditional career outcomes such as academic research faculty. Use of this cohort model resulted in the establishment of groups of trainees who informed our efforts related to culture change through community building and peer mentoring.

Current Emory or Georgia Tech trainees interested in the program applied online. A letter of support from the applicant's PI was required as part of the application process, which

---

BOX 3.1

Specific aims of the Atlanta BEST program proposal

- Aim 1: To expose trainees to a broad variety of career pathways beyond academia.
- Aim 2: To provide trainees deep immersion into specific career pathways beyond academia.

- Aim 3: To better equip faculty at Emory and Georgia Tech to train graduate students and postdoctoral fellows for the 21st century workforce.

---

occasionally forced the initial "difficult conversation" between a trainee and their mentor regarding the trainee's interests in non-academic career options. Since we sought to engage faculty in providing better mentoring around the breadth of career options, having the PI write a letter of support brought them into the cohort of faculty that are committed to this enterprise. The Atlanta BEST Program sought to recruit approximately 30 new trainees each year. We kept this number low so that we could build community and peer-mentoring activities that might have been less effective with larger numbers, and so that we could have a comparison group for evaluation of the effectiveness of our interventions.

Trainee-focused initiatives and outcomes

Atlanta BEST programming focused around five core elements: career exploration, personal development, professional development, leadership and teamwork training, and peer networks. Although we also tested a large variety of interventional programming, created by intentionally leveraging activities already underway at our two institutions, we report

FIG. 3.5 Career tracks explored via the Atlanta BEST Program.

here only those activities that had the greatest positive impact as determined by a survey of our BEST trainees.

Atlanta BEST provided a variety of resources that offered trainees hands-on experiences, informational seminars, and psychosocial support for their career goals. We gathered data on trainee experiences at several points of the Atlanta BEST Program, using entry and exit surveys as well as in-person interviews. In qualitative interviews in which BEST trainees were asked to identify the BEST resources that had been important to their career development and to their development of career-related self-confidence, the most common responses fell into the following general categorizations: 1) resources that offered psychosocial support, 2) reflective resources that provided the opportunity for increased self-awareness, and 3) resources that provided specific career-relevant information (Table 3.1). Psychosocial support-related resources largely came from being members of a cohort and from having a dedicated program manager — who was named frequently and positively — as a resource.

Of particular interest to the BEST leadership was determining whether participation in the program led to increased career self-efficacy for students and postdocs. Prior to the interview described above, trainees responded to a brief online survey asking them to reflect on their confidence prior to enrolling in the BEST Program and at the time of the interview (when they were nearing completion of BEST or had already completed BEST). Trainees were asked questions about the career exploration process, their knowledge and understanding of the type of work and skills needed in various careers, and their overall confidence in their ability to choose a career and succeed in their chosen path. They indicated their current and prior confidence in each item by rating them on a scale of 0—100, with 0 indicating no confidence and 100 indicating maximum confidence. The data show that, upon reflection, students perceive an increase in confidence in each of these areas, including in a number of specific steps along the way to achieving enhanced career self-efficacy (Fig. 3.6).

Graduate students recall that, when they first started BEST, they were the most confident about identifying careers and skills and the least confident about achieving those careers (identifying steps to take, developing networks, etc.). In addition, panel C shows that, reflecting on how they felt at the beginning of BEST, students report having been more confident about excelling in whichever career path they took than they were about actually choosing a path or achieving their career goals. Post-BEST, however, all these metrics were equally high, with mean values above 75. Although these data are self-reflective and were not

TABLE 3.1 Trainees' perceived value of the different types of BEST resources, as indicated by the number of times each type of resource type was mentioned in the survey.

Resource type	Frequency		
	Helpful	Not helpful	Neutral
Reflective processes	34	8	0
Psychosocial support	23	4	2
Specific information resources	20	5	0
Experiential resources	16	2	3

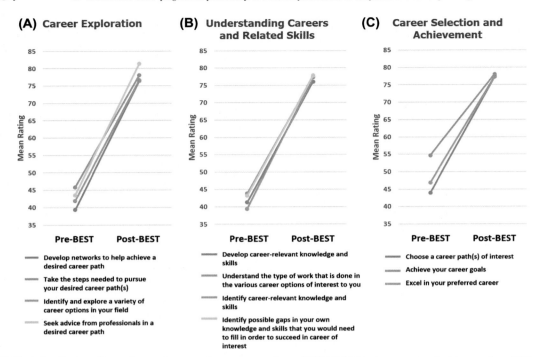

FIG. 3.6   Career efficacy for graduate students, pre- and post-BEST. BEST trainees responded to a series of *"How confident do you feel now about your ability to … ?"* questions to indicate their confidence in different career planning and development aspects on a scale of 0–100, with 0 indicating no confidence and 100 indicating maximum confidence. Reported here are the means of 77 participants' responses.

gathered upon entry, they do show at least some common recognition of gains in these areas. Similar data are available for postdoctoral respondents [3].

Trainees participating in the survey and interviews also were asked to offer specific advice to BEST leadership, and to nominate one or two program features that they thought BEST should absolutely maintain and prioritize as a permanent aspect of the program (Table 3.2). Reflective assessments were of particular importance to the trainees, as were career labs and panel discussions. Notably, cohort Coffee Hours were mentioned in one-quarter of interviews as a priority item since they promote peer accountability.

Data from the surveys also indicate that Atlanta BEST led to the following benefits:

- Improved trainee self-awareness and confidence.
- Greater knowledge of career options.
- Better understanding of trainees' own career interests and choices.
- Increased ability to discuss career interests and goals with others.
- Enhanced agency in career exploration and planning, including:
    - Taking on additional roles and responsibilities inside and outside the research lab.
    - Participating in informational interviews.
    - Attending more activities related to professional development

TABLE 3.2   Trainee advice to Atlanta BEST, obtained through interviews, on what resources to prioritize. Frequency indicates the number of interviewees that mentioned each resource.

Resource	Frequency
Self-assessments (Birkman Method and Strong Interest Inventory self-awareness tools)	21
Career Labs (active workshop)/panels	9
Cohort/Community/Coffee Hours (informal conversations)	7
Resume/CV Workshop	6
Internships (part-time, ~48 contact hours)	6
Networking workshops/events	8
Structural facets (weekly required meetings, organization of workshops)	5

In addition, the surveys of Atlanta BEST participants indicated that the cohort model enabled trainees to identify others with common interests, increased their interdisciplinary peer networks, enhanced their ability to share and discuss experiences, provided opportunity for accountability with peers, and provided safe spaces in which to discuss struggles related to career exploration.

Faculty-focused initiatives

The bulk of this chapter has addressed the needs of pre-doctoral and postdoctoral trainees. However, we know that changing the culture toward accepting and promoting broad career outcomes beyond the professoriate requires shifting the mindset and skills of the training faculty. This can be accomplished, in part, by providing them with tools to improve their mentorship of trainees who seek career options outside of academia. This improved mentorship also benefits trainees who wish to attain traditional faculty careers.

Recent studies have suggested that good mentoring is a crucial component of a trainee's scientific and career development. Like all academic scientists, the faculty depend on the performance and development of students and postdoctoral fellows for the success and growth of their research programs, yet most have not had any formal (or even informal) training in how to be a good mentor. Problems associated with poor or uninformed mentoring invariably lead to unhappy and unproductive trainees, distracting drama, and counterproductive confrontation.

To address some of these issues, it is becoming more common for training grant site visit teams to inquire how training programs help new mentors acquire mentoring skills. With this is mind, the Atlanta BEST Program teamed up with the leaders of the Initiative to Maximize Student Development (IMSD) Program at Emory to develop a group that would focus on and promote various aspects of mentorship: the Atlanta Society of Mentors (ASoM). The Emory chapter of ASoM initiated a pilot, six-session mentoring series in the fall of 2016 for Emory faculty in the STEM disciplines. Sessions included Communications and Setting Expectations, Diversity and Cultural Issues, Conflict Management and Crucial Conversations, Keeping Students on Track, and Promoting Professional Development; two optional sessions were Mentoring and Diversity and Self-awareness and Self-management. Subsequent years included focus topics such as facilitating inclusive environments and exploring culture and diversity.

The faculty mentoring series received excellent evaluations from the participants and is now slated to be offered annually with continued development and improvement. A concurrent, parallel series for trainees also received excellent evaluation from the participants. The participants in this pilot, who were not in BEST, were strongly supportive of the effort to present faculty with tools and improved understanding to enable them to provide improved mentorship. Nearly all faculty participants stated in the series exit survey that they intended to change at least some aspects of their mentorship activities based on what they had learned.

As a consequence of this successful pilot, the major biomedical doctoral programs at Emory voted to require mandatory formal mentor training, such as that provided by the ASoM, for the following categories of faculty:

1. New investigators who joined the Emory graduate faculty within the past two years or who are starting their independent careers and desire to join one of these biomedical programs as training faculty.
2. Existing faculty members who are applying to direct the training of their first graduate student.
3. Investigators who have joined Emory by transfer from another institution where they might have been on the graduate faculty but who have not yet graduated a student.

Leaders of these doctoral programs voted to require faculty to show they have or are in the process of complying with these policies before a student will be assigned to them for long-term mentorship.

Mentoring excellence is a competency to which all of our faculty should aspire. The ASoM workshops are open to all research training faculty, all of whom are highly encouraged to participate. The ASoM is working with the doctoral programs to develop a plan to involve more senior faculty in mentor training, both as facilitators and as active learners. We foresee that enhanced mentoring skills will lead to more motivated, productive, and content students, with the added benefit of allowing faculty to focus their efforts on productive activities rather than on managing conflicts and crises.

Lessons learned: Trainees

Throughout the development and progression of the program, we conducted evaluations to understand and identify the various methods that have had the greatest impact. Formal evaluations included brief, post-event written evaluations; broad electronic surveys of trainees, faculty, and alumni; and trainee interviews. Informal evaluations were mostly focused discussions with small faculty groups, trainees, and student leaders.

Initial evaluations provided early insight into the impact that Atlanta BEST was having on its trainees. Follow-up surveys and interviews brought out specific details regarding the impact of BEST and BEST activities. The evaluations also included a survey of recent alumni from Emory and Georgia Tech to help understand early career issues and to aid in program development.

Like the cohorts of BEST trainees, many new graduates (those within five years of graduating) from relevant programs at Emory and Georgia Tech found themselves rethinking their career paths. Although 36% of alumni had interest in tenure-track research positions at the beginning of their Ph.D. training, only 9% are currently working in or pursuing a tenure-track academic career. By contrast, almost half of recent graduates are working in or pursuing careers in industry (data for BEST participants shown in Fig. 3.7). This is due, in part, to a better understanding of the job market following graduation and to the identification of other careers that utilize their individual skills, interests, and values. Looking back, many of them have stated that, to better prepare themselves for their transition into the workforce, they would have liked to have more professional development opportunities and more skills in core competencies necessary for non-academic careers in addition to the research training they received.

Key findings

Thanks to the various surveys, interviews, evaluations, and outcomes analyses, we have been able to draw the following conclusions:

1. Atlanta BEST has been valuable both for trainees seeking careers outside of academe as well as those interested in the professoriate.
2. Atlanta BEST trainees are independent and resourceful, and are interested in tailored, STEM-specific and career-related support.
3. Resources that can be tailored to each trainee's interests and desired pace are the most valuable resources to them.
4. Trainees greatly value discovering information about themselves.
5. Opportunities for reflection and self-knowledge are critical.
6. Poor advisor support for pursuit of non-academic careers is not insurmountable.

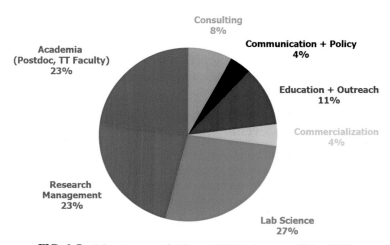

Job Outcomes of Atlanta BEST Trainees

FIG. 3.7   Job outcomes of Atlanta BEST trainees as of May 2017.

7. Program support matters more than advisor support for improving trainee self-efficacy.
8. Trainees value resources that build upon previous learnings.
9. Interest in internships is highly variable.
10. A cohort community is valuable to trainees.
11. Trainees consider access to a dedicated career advisor, mentor, or coach to be critical.

Lessons learned: Faculty

To gauge faculty responses to the implementation of Atlanta BEST, we conducted a survey of Emory and Georgia Tech faculty. Faculty attitudes toward their students' engagement in BEST activities varied widely. Some faculty were concerned that their trainees, paid from their research grants, were spending too much time on professional development and career planning. Other faculty were glad that the institution was supplying this programming, since they knew that they were incapable of doing so on their own. The survey responses are summarized in Box 3.2.

Lessons learned: Specific programming styles and approaches

Over the years, we learned a lot about how properly to execute workshops and events. Here we provide some input on the best practices for us related to a variety of programming types and/or locations.

Cohort-based interventions

Resoundingly, activities/events that actively and explicitly foster a sense of community are impactful. In this context, cohort refers to a single group of trainees who participate in sequential programming at the same time. Successful cohort-based events provided time for activities where participants got to know each other and which facilitated conversations around selected topics. Our cohorts participate in a series of programming and progressive conversations over time. They organically get to know each other outside of the programming, develop relationships, and carry support networks with them during their training. This creates a safe space in which to explore and grow.

Multi-program/multi-campus events

All Atlanta BEST opportunities were open to trainees in biomedically related disciplines at both Emory and Georgia Tech, which are located five miles apart. The variety in scientific backgrounds that is represented — from Biomedical Engineering to Chemistry and more — has really added to the perspectives and conversations within the program. Although trainees had to apply to be accepted into a cohort and thereby received additional exposure to career development, each year we held more open, non-cohort events as we learned what worked. There have been many events where we extended participation to trainees from Georgia State

---

BOX 3.2

Summary of input from faculty groups on professional development programming

Supportive comments

Start with broad topics and career exploration early

- Cycle events annually
- Provide a first-semester course that is department/program specific
- Make resources available to all students; make career exploration and development the new norm

Require elective classes

- Help students choose electives that complement their career interests
- Departments or programs should direct the support for the first-semester required course

Core skills development and relevant resources

- As per conversations with students, increase focus on leadership, interpersonal skills (time management, communication), conflict management, teamwork, etc.
- Help with building relationships/networks

Alumni insight and experience

- Students seem to relate better and pay attention to near peers
- Alumni can provide personal accounts and inside information about daily work activities and environments

Student-led communities

- Senior students can provide well-crafted mentorship
- Faculty or alumni can provide oversight of these groups (e.g., career interest communities)

Have a dedicated space and staff

- Staff can develop an expertise in personal development tools (e.g., Birkman Assessment)
- Staff can serve as career mentors
- The dedicated space can be a safe space, and a centralized location to find staff, hold events, and consult with students

Concerns

Show the value of trainee efforts

- Students appreciate feedback
- Students need to learn how to talk about their professional development activities
- Acquired skills must translate between inside and outside the lab

Limit the requirements

- Students may have too much to do already
- Requirements may not help all trainees
- Provide/allow choices to help the highest number of trainees

Target programming to trainees and make it relevant to them

- Make programming specific and useful
- Help departments/programs organize the programming so they can get trainees' attention
- Avoid interfering with current career development efforts; supplement those efforts

University, University of Georgia, Kennesaw State University, Morehouse School of Medicine, and other surrounding universities. Some Atlanta BEST initiatives collaborated nationally with other BEST institutions, such as University of Chicago, Michigan State University, Cornell University, and Boston University. These connections and networks added value to our programming.

Stage-specific programming

Trainees need different things at different points in their training. There are many things we all want our students to learn while they are on campus, so being thoughtful about *when* certain information is delivered, and in what context, can help make programming more impactful. Two examples are mentoring and career exploration. Getting students to perform well academically and as mentees is the first step to setting them up for research productivity and for more substantive career exploration and development when they are ready. If students are struggling in lab, this impacts their ability to explore careers, to think seriously about their futures, and to make informed decisions.

Focused and active events

Do not try to do too much in one event. Two hours is the ideal length of time for workshops that contain information and activities. Keeping events focused on one to two major objectives, and doing those well, has a better impact than rushing through topics and not discussing them in depth. Leaving white space for discussion is key to understanding, retention, and building community. When possible, those leading workshops can learn and/or get trained in facilitating groups to understand the differences between teaching and facilitating.

Build in activities to engage trainees in the presented topics. Plan for plenty of time for small-group discussions (for example, by breaking a group of 12 into 4 groups of 3) and then bring the discussion back to the big group; trainees benefit from meeting new people, hearing different perspectives, and engaging in smaller conversations. Only allowing for large-group discussions misses the opportunity for these benefits; trainees who are more introverted will prefer and will benefit from small-group discussions the most. If there is a chance to have peers give feedback to each other, take it. Even on topics they do not know well, like preparing a resume or setting up a LinkedIn profile, having trainees review other people's documents helps them learn to give and receive feedback, a skill that should be practiced whenever possible.

Keep events active; Career Labs are a great example of such events. Facilitated by alumni and program staff, these two-hour events are divided so that the first portion (20—30 minutes) is taken up by alumni sharing their background and career path information and the rest of the session is made up of activities that bring representative tasks within the featured career path to life. For example, students in a consulting Career Lab can split into teams and perform a due diligence assessment, while those in a patent law Career Lab can read through patent claims and then write their own. These activities help students do work relevant to their career paths of interest (instead of just hearing about it) and get a better feel for whether or not they would enjoy it in the longer term. If they do enjoy it, the workshop facilitators encourage trainees to undergo further exploration and make new connections. If they do not enjoy it, the trainees are invited to participate in Career Labs that touch on other Ph.D. career options.

In collaboration with InterSECT, an online platform of resources that are essentially virtual Career Labs, the Atlanta BEST Program made several of their Career Labs into InterSECT job simulations. For more examples and online simulations (open to anyone), visit the InterSECT website [4].

Facilitators and presenters

It is important that facilitators or presenters have a good understanding of audiences and their expectations. Some approaches we have found helpful toward ensuring this include:

- Providing presenters with summary descriptions of the type of students to expect (e.g., third year students and above, trainees who are part of a career exploration program, trainees with biomedical sciences backgrounds, shy/reserved trainees, etc.).
- Asking trainees at the beginning of a session (or prior to, such as through an RSVP form) to write down questions so that facilitators can review and address them throughout the session.
- Designing the learning environment and effectively delivering content. Having regular speakers or consistently involving the same people is preferable to having what might appear to be different, random speakers for different topics, even when these facilitators are not presenting a workshop/seminar series. This helps with continuity, with tying all of the material together, and with developing a community of staff and faculty on campus who can get to know the trainees and who can then provide better guidance through continual feedback.
- Tied to the previous point, leveraging other university staff/faculty in broader departments once again helps with the continuity of speakers and with getting to know your program's purpose and audience.
- Leveraging alumni and local professionals. This is a great way to make a session more relevant and impactful.
- Spending time upfront to look at the presentations and discuss the objectives and audience.

Regularity and predictability

Having regularly scheduled events helps trainees plan accordingly. Atlanta BEST trainees met every Thursday from 4:00—6:00 p.m. The trainees have told us that they prefer blocking that time off in their schedules versus having to think about making other times available otherwise. Having a regular schedule also helps avoid the need for trainees to have a conversation with their PI every time they leave lab a little early, since Atlanta BEST is just a regular Thursday event.

Location

Events need to be located close to where students primarily work. When events are on the other side of campus, trainees typically do not attend in the same numbers. We have heard anecdotally that having our events located within the health science center, where many labs are, makes it convenient enough to stop by for an event, then quickly get back to work. Students are also more familiar with that side of campus and have an easier time navigating and determining how long it will take them to get there and back.

Structure and systems

Being explicit about goals and outcomes is a requirement. Worksheets that go along with the delivered content make goals and outcomes more tangible. Giving assignments and requiring deliverables or presentations makes a positive impact for everyone. Give time during the workshops to compete assignments. When trainees complete assignments and share their thoughts, fellow trainees get to hear and see how others approach different topics. They all get much more out of it, instead of being passive and not making time for the material. Be sure to touch on how topics relate to each other; each takeaway is part of the larger system and culture of professional development. This can be easier for programs based on the cohort model.

Recommendations for the future

The goal of Atlanta BEST was to complement the scientific research training of Ph.D. students and postdoctoral fellows at Emory and Georgia Tech by providing them with professional and career development, resulting in more well-rounded graduates who are prepared to succeed in a wide variety of career outcomes. To do this, we proposed providing timely, tested, and integrated resources and programming that enabled trainees to make informed career decisions and develop as professionals. The strategy we proposed to facilitate broad trainee development is supported by benchmarking data from other institutions, from faculty and student input, and from feedback obtained on Atlanta BEST programming and on the graduate school. Accomplishing this goal would require a focused effort, led by dedicated staff with expertise in providing relevant resources and guidance for biomedical trainees, with additional coursework provided by Emory and Georgia Tech faculty supplemented by outside participants. Although our proposal for institutional funding to continue BEST was ultimately not approved, here we present the four major recommendations we made to institutional leaders to accomplish this broad goal.

Recommendation 1

Enhance educational opportunities for trainees in areas that complement the field-specific curricula that they receive via their graduate program. Courses that are offered at peer institutions and that are under development at Emory include the following:

- Project Management
- Drug Discovery & Development
- Commercialization of Technology
- Leadership & Management for Scientists
- Management & Business Principles for Scientists
- Medical Communications

• Business of Biotech

From the BEST Consortium's work to identify best practices, we have the following guidance with respect to this recommendation:

• Experiment; start small, get feedback, and refine.
• Leverage experts even if it requires an investment of resources; everything doesn't have to be "home grown."
• The credibility of instructors matters.
• Leverage your institution's unique circumstances (people, partners, incentive structures, funding).
• Trainees rise to expectations and really want to engage, but not all will go "deep."
• Ph.D. students and postdocs have different needs.
• Small enrollment can mean big impact.
• Institutionalizing courses (making them credit bearing or required) can help with sustainability.

Recommendation 2

Enhance faculty training in mentorship, including the ability to work with trainees toward broad career outcomes both within and outside of academia. This can be accomplished through the efforts of ASoM, described above. We propose that such training should be required of all applicants seeking to join the graduate training faculty if they do not have a record of successful training of pre-doctoral trainees.

Recommendation 3

Centralize the efforts focused on professional development and career exploration for pre-doctoral trainees. This can be accomplished through the establishment of a Center for Graduate Professional Development and Career Planning (or a similar group), which would be charged with working with the local graduate school and directly with the various existing graduate programs to provide programming at the institutional level. The proposed center could also work with postdoctoral fellows.

Recommendation 4

Include stage-appropriate professional development and career exploration as a required component of the graduate curriculum for all Ph.D. students in the natural sciences.

Sustaining the momentum: Cohort-based, stage-specific programming

With five years of funding, the Atlanta BEST Program engaged pre-doctoral and postdoctoral trainees in the biomedical sciences to enhance their ability and confidence to make informed career decisions. Scientific research fundamentals, including critical and independent thinking, scientific reasoning, rigorous research, application of quantitative approaches, and other competencies developed in the lab and guided by the PI provide the

---

BOX 3.3

Professional development and career planning via Grad FORWARD: Programming with context appropriate to each training stage, starting in the first semester of graduate school

	Fall	Spring	Summer
Year 1	**#1 Personal Development** *Self-awareness, Choosing a Mentor*		**#2 Navigating Professional Development** *Leadership Skills*
Year 2	**#3 Mentorship** *Mentor-mentee Relationship*		
Year 3		**#4 Career Exploration** *Explore and Experiment, Academic and Non-academic*	
Years 4+	**#5 Career Strategy**	*Making Informed Decisions*	

---

foundation and core of the Ph.D. training. However, there are broader professional perspectives and skills that also need to be developed, including self-awareness, scientific identity, team science, inclusivity, teaching, and communication of science.

To fill this need, and in order to take advantage of lessons learned from the first four years of funding, Atlanta BEST piloted a program called Grad*FORWARD*, which was centered upon the delivery of stage-appropriate programming intended to enhance student success during graduate school and to enhance their ability to make informed career choices leading to long-term career success (Box 3.3). The program was designed to begin in the first year of graduate school and to end roughly in year 5, with each group of students entering graduate school in the biomedical sciences undertaking the program as a cohort. Major topics covered included those shown in Box 3.4.

The eight biological/biomedical doctoral programs at Emory voted to require all entering students in these doctoral programs to participate in Grad*FORWARD*. This programming was intended to become an integral part of graduate training at Emory, irrespective of a student's intended career path. The expectation was that the professional development skills and career exploration opportunities provided by Grad*FORWARD*, in addition to enhanced preparation for academic careers, would help students become confident, capable, and competitive for the next steps after graduating.

Conclusions and predicted outcomes

In summary, the lessons learned from Atlanta BEST will transform Ph.D. graduate student training by complementing the excellent scientific research training at Emory and Georgia Tech

---

BOX 3.4

Key concepts for each stage of GradFORWARD

#1 Personal Development	Self-awareness, defining success, building networks, understanding culture, addressing wellness
#2 Navigating Professional Development	Resilience, time management, project management, people management, conflict management
#3 Mentorship	Facilitating mentorship development, situational mentorship, expectations, conflict management, lab culture
#4 Career Exploration	Mindset and strengths related to work environments, career paths, building a career narrative, alumni roundtable, experiential activities
#5 Career Strategy	Designing a strategy, producing quality career documents, interviewing and negotiating, networking with practicing professionals

---

with integrated professional development programming and resources that enable students to explore and prepare themselves for their future career success. The vision for the future entails a shift in the model of Ph.D. training from one that is PI-centric, training primarily for a single career path as academic faculty engaged in research and education, to one of shared roles where students feel enabled to explore their career options and interests and where the institution provides resources and programming to fill in the gaps in skills and knowledge (Fig. 3.8).

Building upon existing institutional efforts and our experiences after five years of implementing the Atlanta BEST Program, we identified the programming and approaches that would likely have the greatest impact on graduate students' ability and confidence to make informed career decisions regarding their future. While the scientific research knowledge, skills, and competencies developed in the lab and guided by the PI provide the foundation and core of Ph.D. training, as always, the proposed professional development skills and career exploration opportunities would help students become confident, capable, and competitive for the next steps after graduating. Creating centralized resources and programming for Ph.D. students, organized by a dedicated staff, will ensure that graduate students complement the excellent scientific research training that they already receive with an enhanced ability to explore different career options and develop the skills and competencies necessary for their academic or non-academic career futures.

The intended trainee outcomes are:

- Being informed about career options that leverage Ph.D. training.
- Becoming confident, capable, and excited about their next steps after training.
- Being competitive and prepared for the jobs they want.
- Having eventual long-term career satisfaction.

The intended institutional outcomes are:

- Spreading the word of institutional excellence.
- Maintaining Emory's and Georgia Tech's strong reputations.

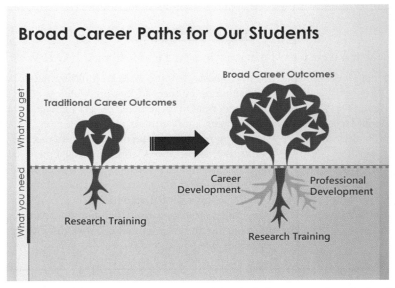

FIG. 3.8   Predicted outcomes of shifting the training model to include career and professional development in addition to research training.

- Further developing and strengthening alumni networks.
- Enhancing faculty and student recruitment and retention.
- Expanding the institutions' positive impacts on the national and global stages.

Changing the culture around Ph.D. training will take resources in the form of staff, infrastructure, and programming to ensure the institution's future ability to compete for and recruit quality students and to enhance our reputation as leaders in research and educational excellence. These are important steps toward broadening the career paths for our biomedical trainees (Fig. 3.8).

Acknowledgments

The Atlanta BEST Program was supported by NIH 7-DE024096. The program also received support from the Laney Graduate School at Emory University. The authors declare no competing interests.

References

[1] National Institutes of Health. Biomedical research workforce working group report. Bethesda, MD, USA: National Institutes of Health; 2012.
[2] National Institutes of Health. NIH director's biomedical workforce innovation award: broadening experiences in scientific training (BEST) funding opportunity announcement RFA-RM-12−022. Bethesda, MD, USA: National Institutes of Health; 2013.
[3] St Clair R, Hutto T, MacBeth C, Newstetter W, McCarty NA, Melkers J. The "New Normal": adapting doctoral trainee career preparation for broad career paths in science. PLoS One 2017;12:e0177035.
[4] InterSECT website. Available from: https://intersectjobsims.com.

Michigan State University BEST: Lessons learned

Stephanie W. Watts, Karen Klomparens, Julie W. Rojewski

The Graduate School, Michigan State University, East Lansing, MI, United States

Cast of characters

Stephanie Watts, Ph.D.: Biomedical faculty member in the Department of Pharmacology & Toxicology and an Assistant Dean of the Graduate School at Michigan State University (MSU) when this story began. Committed to training biomedical scientists.

Karen Klomparens, Ph.D.: Dean of the Graduate School at MSU when this story began, Dean Emeritus as of 2015. A plant biologist, she is savvy about MSU and multi-institutional projects.

(Continued)

BEST: Implementing Career Development Activities for Biomedical Research Trainees
https://doi.org/10.1016/B978-0-12-820759-8.00004-8

Kevin Ford, Ph.D.: Organizational psychologist at MSU with expertise in evaluating institutional climate and climate change. Conductor of Windrose surveys and internal MSU surveys.

Trish Labosky, Ph.D.: NIH Program Officer for BEST, Program Leader in the NIH Common Fund.

Doug Gage, Ph.D.: Assistant Vice President in the Office of Research and Graduate Studies of MSU. A biomedical scientist.

Julie Rojewski, Ph.D.: MSU BEST Manager.

Rich Schwartz, Ph.D.: Associate Dean for Graduate Studies in the College of Natural Science. A biomedical scientist.

Katy Colbry, Ph.D.: Assistant Dean for Graduate Student Services in the College of Engineering.

(Continued)

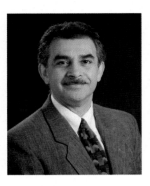

Manooch Koochesfahani, Ph.D.: Associate Dean for Graduate Studies and Faculty Development in the College of Engineering.

John LaPres, Ph.D.: Professor of Biochemistry and Molecular Biology and head of the BioMolecular Science Graduate Gateway Program.

Julia McAnallen, Ph.D.: Inaugural director of the Ph.D. Career Placement Office at MSU.

Act I: Scene I

Late winter 2012/early spring 2013

SETTING	*The office of Dean KAREN KLOMPARENS, Dean of the Graduate School in Chittenden Hall at Michigan State University. There are books and papers everywhere; typical for an active and busy academic.*
AT RISE	*KAREN and STEPHANIE WATTS sit at the table, KAREN has a notepad; STEPHANIE has an iPad.*

Narrator: The Graduate School at Michigan State University, led by Dean Karen Klomparens from 1997 until 2015, has been a leader in a number of areas of graduate research. MSU was one of the first universities to offer career and professional development uniquely tailored to Ph.D. students and postdocs. It developed tools like the PREP model for graduate education and web resources like "Career Success," which is used by campuses around the country to equip students and their mentors with the tools they need to support trainees' growth throughout and beyond their doctoral experiences. Both have been shared and implemented widely.

At this time, the MSU Graduate School offered workshops for professional development and supported the "Graduate Life and Wellness" program, which recognized that graduate students need physical, emotional, and psychological wellness to succeed in graduate school and in their careers. These pieces provided a foundation for MSU BEST.

Scholarly research in graduate education, through federal (NSF AGEP, Council of Graduate School Projects) grant- and cohort-driven projects, has long been central to the MSU Graduate School. The Graduate School has also been the place for the convergence of postdoctoral fellows and underrepresented populations, neither of which is numerous at MSU.

Stephanie: Karen, I am so excited to talk to you! Some of us in Pharmacology & Toxicology just watched an NIH Director's Fund webinar. There's a new request for applications, and I think we would be really well positioned to apply for a new program called BEST, Broadening Experiences in Scientific Training. The overarching goal seems to be to run our own experiment to determine what interventions work best to help biomedical trainees prepare for careers in many arenas. Sounds like we'd eventually compare, contrast, and consolidate with other schools that do this BEST experiment.

Karen: Tell me more about the project and why you think we should go for it.

Stephanie: Well, the obvious answer is that we already have so much in place at the Grad School upon which BEST could be built. And this is an experiment! We can construct a BEST program that uses what we already have and run a test within it to find out what interventions help trainees find their professional path. We're not starting from scratch. You know I love visualizing things, so look at this:

(Stephanie pulls out her iPad and shows Karen Fig. 4.1 *on it).*

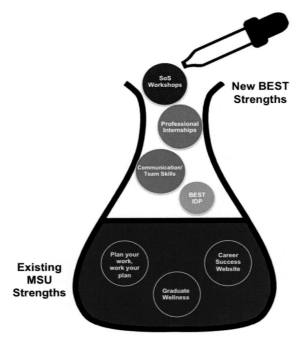

FIG. 4.1   Cartoon of what MSU's career development program could be if MSU participated in BEST.

Our Biomedical Science Program is ideally suited to this, and the Engineering program is doing more and more biomedical work with the new Biomedical Engineering program and the Institute for Quantitative Biology. So, we have more people than ever doing biomedical research on campus.

We know from the growth of the Office of Ph.D. Career Services that students are hungry for information and training in "alternative careers." And enough has been discussed in professional meetings that faculty are also coming around to recognize the importance of training for multiple career tracks.

Karen, laughing: It's not an "alternative" when most of the students end up in such careers. Julia calls them "expanded careers," so we can use that term instead. But from what you're saying, I'm encouraged.

Stephanie: That's great! And I want you on board *now* because it's very important to me that we come into this experiment intending for this to continue within MSU on the long term. With your support, we have the chance to really build something that lasts.

It's the perfect opportunity to build on what we have, to design and run an actual experiment about what *works* in professional development — we can test it out! — and then keep what works into the future.

Karen: Yes, but before you write an application, do some work in gathering your stake-holders, both on and off campus, to help this be successful. Faculty absolutely have to be there. You will want to include graduate directors, department chairs, and associate deans for research. Ideally, the Office of the Vice President for Research and Graduate Studies would be a partner. As I think of it, you should probably contact Doug Gage. All of these individuals and their offices have a dedicated piece in the success of graduate students and postdoctoral fellows, so engage them sooner rather than later.

You'll also need a central group of people who could help extend your reach through campus and carry out the experimental part of this grant. You'll want good assessments and evaluations of the current program. I'd contact Kevin Ford in the Department of Industrial/Organizational Psychology [I/O Psych], Manooch Koochesfahani in the College of Engineering, and Rich Schwartz in the College of Natural Science to see if they'd be interested in partnering. The Grad School uses faculty in I/O Psych routinely for program evaluations because we get a serious formative and summative evaluation that includes both qualitative and quantitative data. They are independent of the Grad School and do a great job of holding our feet to the fire and giving us pointed, evidence-based data and interpretations.

Stephanie: OK. Those are great places to start. I'm excited you're on board … Now the real work can begin.

(BLACKOUT)

(END OF SCENE)

Act I: Scene II

Spring 2013

SETTING	*A conference room with a large conference table.*
AT RISE	*STEPHANIE, KEVIN FORD, KAREN, RICH SCHWARTZ, JOHN LAPRES, MANOOCH KOOCHESFAHANI, and KATY COLBRY are sitting around the table, with an open seat downstage, so the audience sits at that "head" of the table. Upstage, at the head of the table opposite the audience, is a large screen upon which various images are projected.*

Stephanie: Thank you all for meeting! I'm excited that you're interested in this project, or at least interested enough to come to this planning session.

I know you've each already heard me talk about the BEST program and NIH's goals for it. I have met with most of you individually, as I've tried to meet with as many stakeholders on campus as possible. I've had a series of meetings with graduate directors and administrators, primarily research deans and chairs, of our biomedical departments. That was most of spring 2013.

I should say that not everyone I spoke with was on board. Many expressed the reservation that this could be a tough sell for the students and their faculty mentors if the students really had to spend significant time outside of the laboratory to do MSU BEST. Or they felt that it wasn't really "our job" to offer job training. That said, MSU has some folks known for their great mentoring, and they are all for it. So, we are moving ahead with the experiment, while actively listening to those in doubt to help us understand how we might get them interested later.

I will say that is really helpful, as I am getting an appreciation for these early concerns and reservations; I respect this.

Kevin: Plus, I think we have an opportunity to study these concerns. As an organizational psychologist and evaluator on grants like these, I hear that kind of feedback and realize we can run a study on this to capture faculty beliefs. Faculty rely on data for research, so this is an opportunity to provide both short- and long-term data for them.

Karen: That's why it is so, so important to have an internal evaluator — and a non-biomedical scientist! — involved. We need this objective, evidence-based view throughout. And we know that Kevin will help us by asking difficult questions.

Kevin: We should determine *experimentally* whether the interventions that trainees undergo are effective in guiding them to a career in which they are satisfied. It makes sense to think of this in terms of short-, mid-, and long-term goals for this project. Something like this, which I know you've seen. I appreciate all the feedback you gave me to help build this.

(Fig. 4.2 *appears on the screen, and everyone looks at it*)

Stephanie: Thanks, Kevin. We can revisit this, of course, but as a guiding framework, it provides us a useful way to think about our program because it recognizes that trainees are probably looking for different levels of engagement — low, medium, and high — and would be engaging with the program at various times throughout their training.

This brings us to another question: how do we actually do this? We've talked about proposing that MSU BEST be based on a cohort model. This means we'd deliberately select students to participate through an application process, and then offer them an experience they'll go through together.

Karen: MSU really does have a history of running programs in cohorts or on the same time frame. Look at our cohort-based teaching programs like FAST, the Future Academic Scholars in Teaching, or the Leadership Fellows. Our internal data supports that students learn well in this structure, and the research on peer-to-peer learning and social support applies here. I'd suggest using the cohort model — we'd be able to do rigorous, formative, and summative evaluations this way.

We should limit the number of trainees that can participate in any one cohort, mainly because we are currently unsure of the number of internship/externship opportunities

Short Term	Medium Term	Long Term
Affective •Openness to investigating non-academic careers •Openness to new ideas and training	**Affective** •Improving self confidence •Openness to the concept of wellness	**Affective** •Climate change in faculty accepting parallel mentoring •Culture change in support of non-academic careers •Improved trainee self confidence
Knowledge •Greater knowledge of MSU Career Services	**Knowledge** •Greater knowledge of what employers want skill-wise •How to work in a team •Knowledge of a career plan	**Knowledge** •Greater knowledge of career options for trainees •Knowledge of how to obtain a job
Skills •Self analysis	**Skills** •How to work in a team •How to develop an IDP	**Skills** •Improved oral communication (lay public, private sector) •Higher quality resume
Behavior •Completing Orientation, Career Success	**Behavior** •Completing MSU BEST IDP •Completing Internships •Practicing Wellness	**Behavior** •Proactive in search career opportunities •Choice of pursuing biomedical career made earlier
Outcomes •Website for MSU BEST	**Outcomes** •Completed internships	**Outcomes** •Decrease time in postdoctoral fellowship •Expand type of jobs obtained; success in obtaining jobs •Elevate the diversity of demographics in programs •Improve recruitment of strong students to MSU •Increase retention rate •Decrease time to degree

FIG. 4.2 Short- (≤1 year), medium- (≤3 years), and long-term (≥5 years) goals of MSU BEST relative to trainee outcomes.

we could have for our trainees, which is definitely an element we plan to build into this experiment. So, there are two reasons for choosing a cohort model. First, our experiences show that it works with graduates and postdoctoral fellows. Second, it's practical!

Kevin: Can we talk a little more specifically about the internship/externship experiences?

Stephanie: Sure. We think it's really important for trainees to have some hands-on experiences, something different from outside the lab. We want them to see and really try on different contexts in which they might "do science" as professionals.

Trainees are used to doing rotations in a biomedical lab as they choose their place to perform their Ph.D. work. In doing this, they test out a type of science, a mentor, and a lab group. That said, the idea of testing out a profession — a "rotation in a profession," if you will — is brand new to biomedical disciplines at MSU. We already know that some folks are concerned about how much time out of the lab BEST activities will require, but they are even more concerned about how much time an internship or externship will take.

Karen: Exactly. We plan to have all trainees do internships, but we know that these will be limited because of our location in East Lansing, which is not rich in the biomedical industries. We're not in New Jersey, or Boston, or the Bay Area. We don't have that many potential partners, so we have to be realistic about how many opportunities we can help build. As such, keeping the cohort small will actually maximize the number of trainees we can realistically reach. That's a very real disadvantage at a place the size of MSU, sure, but I think it's the right way to go. Part of our philosophy in the Grad School has been "adapt, don't just adopt" someone else's model — see what ideas we can borrow from others and adapt them to fit MSU. We'd hope that other grad schools who would want to use our career and professional development programs do the same.

Stephanie: What other challenges would a cohort model present that a *cafeteria* or an *à la carte* model would not?

Karen: For the cohort model, you'd need a group of trainees (and their faculty mentors) who would *stay* engaged through the experiment. These are trainees who would not just pick and choose when to attend BEST activities like they could do with an *à la carte* model, but rather, they would do the very best they could to come to as much as possible with their fellow cohort members.

This also gets to the issue of the "dose" of BEST. We are betting that a high dose of BEST activities will do the best by trainees; the cohort model allows you to do this. What this model lacks is the ability to reach a lot of people. It's also expensive on a per person basis because we would be reaching fewer people through more intensive experiences.

Even so, I still think it's the right approach for MSU.

Stephanie: Let's write the grant, then!

Fade to black   STEPHANIE, KEVIN, and KAREN *are hunkered over the table, writing on their laptops and exchanging papers.*

(BLACKOUT)

(END OF SCENE)

Act I: Scene III

Fall 2013

SETTING	*Conference room with large table. Stage is dark, with characters sitting still around table.*
AT RISE	*STEPHANIE, downstage right, looking at her laptop. She lowers her head with a disappointed expression; she slowly reaches for her cell phone. Lit by a spot.*
	TRISH LABOSKY sits downstage left, at a simple desk, with a phone. Lit by a spot.

Stephanie: Dr. Labosky, thanks for taking my call and for letting us apply to BEST. I just got the rejection. As you know, the MSU team spent spring of 2013 writing the grant, and we got good feedback. I think we did a good job addressing concerns of the reviewers, so I'm disappointed we didn't make this first cut.

But do I understand correctly that there will be a second RFA? Can MSU still apply? We were told that our application was "too academic," and I'm having a difficult time understanding what that means. Too needy? Not enough in the social science realm? How does this basic biomedical scientist interpret this?

Trish: Yes, there will be a second offering, and yes, you can — and should — apply. I can't say specifically what the issues were with the proposal...

Stephanie: At least some of the feedback suggested our application would be strengthened if we revised the proposal so that it was less like a basic science research grant. I think bringing in fresh eyes from outside biomedical science could help.

Trish: That sounds very doable. Go for it!

(Stephanie hangs up the phone, thinks for a second, and then picks it up again. She dials and waits for a minute for the phone to ring on the other end)

Stephanie: Hi, Katy? This is Stephanie Watts. I have a project, and I think you're just the person for the job.

(FADE TO BLACK on STEPHANIE and TRISH. Light then shines on a big group of characters: KATY, RICH, KAREN, MANOOCH, JULIA MCANALLEN, and JOHN LAPRES. They are sitting around a conference table that is littered with laptops, papers, books, journals, etc. Silently, they gesture as if talking with one another. On the screen upstage, at the head of the table, you can see the BEST proposal being edited by the group in real time as they silently edit together.)

(FADE TO GRAY, SPOTLIGHT ON NARRATOR)

Narrator: In the spring and early summer of 2014, Katy Colbry of the MSU College of Engineering took our application and turned it upside down in presentation style, making it highly organized and streamlined. During this time, the Graduate School formally opened the Ph.D. Career Placement Office, headed by Julia MacAnallen, Ph.D.

The final grant group was complete: Katy and Julia joined Kevin Ford, the organizational psychologist; Rich Schwartz, Senior Associate Dean for Research in the College of Natural Sciences; Karen Klomparens, Dean of the Graduate School; Manooch Koochesfahani, Associate Dean for Graduate Studies in the College of Engineering; and John LaPres, Director of the Biomedical Sciences umbrella program. Each one had been engaged in the BEST project from the beginning. (*Each actor looks out to the audience as his or her name is called.*)

With budget assistance from Shobha Ramanand of the MSU Central Office of Planning and Budgets, the MSU BEST team resubmitted the grant, which showed widespread institutional support for the program. In August 2014, the NIH asked for clarifications and revisions, which the MSU BEST team provided.

In September 2014, Principal Investigator Stephanie Watts had just boarded a plane to a conference when she received the call that MSU was awarded a BEST grant in the second round. She shocked her fellow passengers when she proceeded to do a happy dance in the aisle.

(*Spotlight shines on Stephanie, who yells out and does a happy dance.*)

(BLACKOUT)

(END OF SCENE)

Act II: Scene I

Early September 2014

SETTING	*DOUG GAGE's office in an administration building on the MSU campus, at the Office of the Vice President for research and graduate studies.*
AT RISE	*STEPHANIE and DOUG sit together. STEPHANIE has her iPad; DOUG has a notepad and stylus.*

Stephanie: We are incredibly fortunate to be one of 17 groups who get to do this BEST experiment, and I'm excited to talk to you about it. We proposed a cohort program for biomedical science and engineering graduate students and postdocs, with a focus on communication, wellness, and teamwork. We will require *two* externships of each trainee.

What we're *unsure* about is this: exactly what kind of time will BEST take for our trainees? One of the things we are most worried about is that faculty will not be willing to allow their trainees to do BEST because of the amount of time outside of the lab it requires; we've been hearing this since we started considering submitting this grant. We don't have data to support or refute this idea yet, but when we talk about BEST with PIs and other faculty members, *time* is always the part that worries people. They are worried about the time their trainees will need. They're also concerned about how much of their own lab's work will get done if their trainees are off doing BEST.

So, I have been trying to think of ways to soothe some of these concerns. How about this: would it be possible to designate a mini-grant to the PIs of BEST trainees who are willing to do this experiment with us? These dollars would be a "thank you" to them for being willing to take a chance on MSU BEST, and would offer support that they can then decide how to use. They could use it to cover the expenses that stem from their trainees being gone and could hire assistants, purchase materials, or whatever else they feel is appropriate. They could use it to support professional development for their trainees. They would get to decide.

Doug: Yes. BEST is exactly what we should be doing, and I am supportive of both the idea of providing training and of the solutions you have in mind to overcome resistance from faculty mentors and PIs. You know that I'm a plant biochemist by training, and yet I've been a faculty member, have worked in big pharma to research and develop new drugs to treat human disease, and now I'm Assistant Vice President for Research and Graduate Studies! I'm exactly the kind of person who is doing a job that is different from what I was formally trained to do.

So, here's what we can do. For each year of BEST, our office will give you $2,000 from our MSU general fund budget to give to each PI of trainees involved in these additional opportunities in recognition of their support and of the fact that the process will take their trainees out of the lab and that could, but hopefully will not, negatively affect their laboratories. We believe in what you are doing and will use our own budget to indicate that support.

(BLACKOUT)

(END OF SCENE)

Act II: Scene II

Early September 2014

SETTING:	*Back in the office of KAREN, Chittenden Hall, MSU campus.*
AT RISE	*STEPHANIE, with her laptop, sits across the table or desk from KAREN.*

Stephanie: Well, here we go! We got the BEST funding and an opportunity to do this experiment. I'm so excited! How the HECK do we really start this?

We learned about this award in September, and it's October now. Can we practically start with our first cohort this year, 2014? This biomedical scientist is feeling somewhat out of her comfort zone with what to do first! I prepared a flyer to announce this program (Fig. 4.3 *appears on the large screen, center upstage*), but how do I make this real? This is *nothing* like an experiment I can run in the lab.

Who: Biomedical and Engineering PhD students and Postdoctoral Fellows

Why: To empower biomedical/engineering trainees and their mentors with career skills that maximize possible career choices

How: Bimonthly workshops to learn about different biomedical career pathways and develop professional skills in communication, teamwork and wellness in year 1; Work in consultation with your mentor to develop and IDP and explore two or more externship career opportunities that match your interests and goals in years 2 and beyond.

To learn more, or to apply to become part of the MSU BEST program, see www.best.msu.edu.

FIG. 4.3　Flyers for BEST used from 2014 onward, with some modifications.

Karen: Go to your stakeholders and those individuals who are most willing to try this experiment. Explain to them that NIH BEST is now here and something their trainees can use. You staying in touch with faculty mentors is going to be essential. Start with the people you know who already believe in this program. Get this first group together pronto and MOVE.

Then we can talk about expanding your reach: off the top of my head, I can see you going to individual department meetings to introduce BEST, for example.

Stephanie: Do I go for everyone?

Karen: I don't think so. Not right now. All of the BioMolecular Science (BMS) programs should be involved, though. Who do you know who is well recognized as a thoughtful and active mentor? Ask them to have their graduate students join now, as part of the start of this experiment. We need a few key early adopters.

Stephanie: Even though it's truly an experiment? Even though they'll be the guinea pigs?

Karen: They are all scientists. The faculty and their students will see this is valuable. They will see a good experiment and want to be part of it. Start there. I think you'll be surprised.

I know it would be ideal to have someone to help you get started, but we know that takes a while. You accounted for a Program Manager in the budget. Have you thought about who might fit that role?

Stephanie: A little, but I definitely want to rely on you. This is an administrative position, which is different from other positions I have hired for before. Even my lab managers have to be scientists first, albeit ones with excellent organizational skills.

Karen: You can go in many different directions. I know of one common way that is exactly what you said: to prioritize the scientific training and look for someone with organizational skills. I think we should house this BEST Program Manager in the Grad School itself. That way, they will have a whole set of mentors and idea-people to help out daily.

Stephanie: Remember how we're part of the second group of awardees? I think that's the approach some of the first group of BEST awardees took. There is definitely value in having someone with a biomedical background who really understands the culture of science and knows what our trainees are up against. Such a person would also know what skills are needed to succeed but aren't currently part of biomedical training. That makes sense to me.

Karen: I get that. Plus, it represents an example of a "non-PI" job for a biomedical scientist, exactly what the BEST grant is looking to explore.

But I think there are other ways of looking at it, too. There are people with training in professional or faculty development. There are people who are education experts — who know how people learn — and there are people who are higher education experts — who know a lot about university contexts, program management and evaluation, and such. There are organization people: organizational psychologists, business people, etc. They wouldn't know about the science, per se, but they would know about people and organizations. That is another approach.

Stephanie: Do you think one way is better than another?

Karen: Well, obviously, you want to put together a search committee. They will have the final say on any sort of job description. I know I say this all the time, but it's a really important point: there is far more value in extending your network and getting more diverse viewpoints than there is in getting more people who think and were trained like you. Any time you have to work with people who don't already do things like you do, to draw from a diverse pool of expertise and experience — it's generally better, especially for a campus-wide project.

But we're getting ahead of ourselves. You need to go find yourself some trainees for this grand experiment!

(FADE TO GRAY, SPOTLIGHT ON NARRATOR)

Narrator: In December 2014, Stephanie Watts, PI of MSU BEST, convened the first meeting of trainees, 21 strong.

At this first meeting, she introduced the "BEST Promise," a document that was signed by both the trainee and mentor verifying that both knew the trainee was in BEST and the mentor was prepared to be supportive of the activities of BEST. It was important to the MSU team that participation was open: trainees were to have explicit support from their mentors and no one was to hide anything from their PIs.

The upper administration of Michigan State has been supportive from the very beginning of MSU BEST. The Office of the Provost and the Office of the Vice President for Research and Graduate Studies collectively committed to providing a $2,000 stipend for the PI of each BEST trainee. This meant that university leaders were not only aware of BEST from its earliest days, but they also had formal buy-in to the program and were invested in seeing it succeed. This early support proved critical in the later institutionalization of the program and in getting a commitment from MSU that BEST would continue after the grant funding period was over.

For the first group of students, each PI got their stipend at the beginning of the program. The MSU BEST team later changed the distribution model so that the stipends were given to PIs when the student started their first externship after a full year of BEST participation; this maximized commitment on the part of trainees and helped to emphasize that the stipend was to recognize the time required for these short-term internships. In one case, a trainee dropped out of the program after the PI had already been awarded the stipend. The PI was asked to return the money and he agreed to do so.

Stephanie and Julia also introduced the pillars of MSU BEST (Fig. 4.4); the Spheres of Success, internship areas they knew could be provided (Fig. 4.5); and a time frame for how they anticipated MSU BEST trainees to proceed through this program (Fig. 4.6). With the help of KINDEA Labs [1], they also created an animation [2] to help explain what MSU BEST really was within two short minutes.

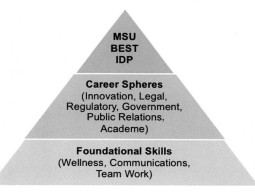

FIG. 4.4 Foundational skills that are pillars of BEST (bottom), career spheres (middle), and a capstone (top) of MSU BEST Individual Development Program, using the myIDP web platform. These skills and experiences must be fulfilled before trainees participate in internships.

FIG. 4.5 Spheres of Success (SoS). MSU BEST developed this model to reflect the five general fields of professional opportunities that we envisioned would appeal the most to trainees and in which we could support internship/externship experiences. At the center, we refer to workshops and other trainings in transferrable skills that support success in all of these fields.

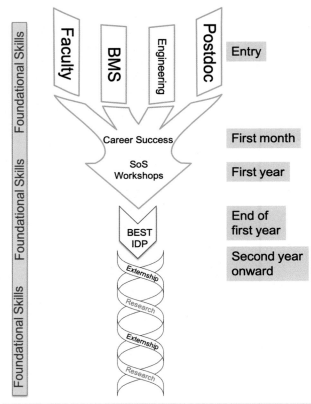

Goals: Ph.D. Defense, Job Placement, Faculty Buy-in

FIG. 4.6 Shared timeline for the progress of "BESTies," trainees who participate in BEST, from entry through an anticipated second year during which they would take part in an internship/externship.

(*BLACKOUT*)

(*END OF SCENE*)

Act II: Scene III

September 2015

SETTING   *Conference room. Diet soft drinks and chocolates are present.*

AT RISE   *STEPHANIE and JULIE ROJEWSKI sit together.*

Stephanie: We've come so far, haven't we?

Julie: It's been a whirlwind, definitely.

Stephanie: That's a nice way of putting it. You came on board as our BEST Program Manager at the beginning of the year after we had already gotten started with the first cohort, and yet you were able to jump right in, which I am so grateful for.

Julie: Thanks. I think it helps that I had experience doing a lot of the things you already needed from the get-go. Managing big grants, program planning and evaluation, and a lot of professional development programming and training.

Stephanie: And your degree in higher ed brings in a different perspective. I appreciate that, too!

Julie (laughing): My *almost* degree, right? I'll get there.

Stephanie: Of course you will! And you have a ton of experience in this world already. But I appreciate that we have such different views. You, a social scientist with a humanities background; me, a hardcore scientist.

Julie: And remember, we did the Myers − Briggs Type Indicator (MBTI®) personality tool, and we were polar opposites there, too!

Stephanie: But now we're in BEST together, and, as we've seen, a diversity of views and honest questions can help move a project forward. So, let's figure out what we've done and where we need to head.

Julie: Well, the first thing we did was agree that I could call our trainees BESTies.

Stephanie (laughing): I think the trainees love and hate it.

Julie: I love it. They are warming to it, even if they groan a bit.

But more concretely, in terms of our accomplishments … Our program is based on getting our individual cohorts together for professional activities. We've already done things like MBTI® testing, resilience training, lots of networking, CV/resume, leadership and teamwork − both the more theoretical aspects and the practical skills they can use to be good teammates and effective leaders − and, of course, science communication and networking. And other topics, too, because as you know, my goal is to provide BESTies with several different types of experiences and skills, including ones where they get to know themselves better, get some clarity about their goals, and more.

These activities are well received by our trainees. We know this from our workshop evaluations. The curriculum is generally going well, and I appreciate how open you are to letting me use my professional and educational background to exploring and innovating here.

I also think we have been very smart in always holding our activities in the evenings so that we don't interfere with lab or teaching work.

There is one type of activity that is not doing too well, though: the ones that are more social or team building-centric in nature. We built those activities in as a way to build cohesion in the cohort, as a way for BESTies to have fun together. We've tried a variety of things: mindful meditation, ballroom dancing, knockerball, picnics... but they're just not participating.

Stephanie: I'm betting that trainees think of MSU BEST work as a strictly professional activity, not a social one. Their "off time" is really their "off time," and we cut into their evenings as it is with our events.

Julie: That's a good point.

Stephanie: We've learned the hard way that, at least as a cohort, social events don't fly very well. I am hopeful, however, that when we pull all our cohorts back together at the end of this grant, we will have an excellent turnout and will get to see these folks socially.

Julie: But in the meantime, I think we're in agreement that we need to abandon this part. It's not worth our resources or time.

Stephanie: Agreed. We tried.

Another thing I'm glad about is the fact that we expanded our original 7–8 departments to over 18 departments at MSU that do biomedical research. This has allowed us to get the message of MSU BEST across our campus and to grow our cohorts to the near 30 students that we are comfortable helping each year. We still have an eye on continuing the program after the grant and expanding our reach.

Julie, what else has been a challenge? Could we have foreseen these?

Julie: Internships and externships! I don't think we expected this to be the challenge it has been, or at least not for the reasons why it has been a challenge.

Different offices and businesses in and around MSU and even further out in the state have been remarkable in answering the call to give a biomedical trainee an externship that lets them experience a career. Most have been eager to work something out when a BESTie approaches them.

But, somewhere along the way, a lot of trainees got the idea that we were just going to hand them an internship/externship. They would come to me and say, "I want to work in industry, so I'd like to have an internship in pharma. What do you have for me to pick from?"

And that's just... not how it works. Even if we did have that — which would effectively be a co-op model that other BEST universities DO have — we'd have to be able to promise our partners that we can have a qualified and committed BESTie to fill each slot, every single time. As it is, even when we have amazing partners come to *us* and say they'd love someone to intern with them and I offer it up as a lead, trainees just aren't biting.

Stephanie: That's not what we planned for.

Julie: Exactly. The result is that I have to spend a fair bit of time explaining to BESTies that ideally, after all of these workshops, they have the insights about themselves and their goals, the active support of their PI and others, and the skills in networking to be able to reach out and start a conversation with someone they know, or someone they can get connected to. We can *help* them think through the steps, but we never intended to just slot people in places.

Stephanie: And we've talked about how this process has been paralyzing for some trainees. They just *won't* do it.

Julie: It has been a challenge! But in other cases, the process has been liberating. Many trainees have identified incredibly interesting places to serve. I'm thinking particularly of one neuroscience trainee who does work on depression who worked her network to find a clinician at a mental health facility that works with pharma companies to run clinical trials. She has connected with their medical team and gets to see so many relevant steps: how they get connected with pharma, how the protocols are developed and approved, and how the trials are discussed among the medical doctors, researchers, and regulators. I couldn't have imagined such an opportunity on my own, but she and her PI worked together to mine their networks with her goals in mind and found an excellent collaborator for her internship/externship. It has been a fantastic experience for her.

Stephanie: And I'm thinking of one trainee who decided she was interested in patent law. I know she had already been thinking of becoming a certified patent agent, but one panel discussion with lawyers, two legal-related internships, and one successful dissertation later, and she's on a full ride to law school. She will be the first to say that wouldn't have happened without BEST, and without us nudging her out of her comfort zone to try something new. Now that? That I am so proud of. We helped make that happen.

Spotlight fades on Julie and Stephanie as they strategize, laugh, and work on this program that they love.

(BLACKOUT)

(END OF SCENE)

Act II: Scene IV

September 2015 through August 2018

SETTING	*Conference room.*
AT RISE	*STEPHANIE, JULIE, and KEVIN sit together.*

Narrator: MSU BEST took its mission that this needed to be an experiment seriously. From early on and throughout the process, they recognized the challenge of collecting some of the data they wanted to track.

MSU participated in the rigorous cross-site analysis that generated powerful data across all 17 BEST campuses. But the MSU team realized there were additional research questions they wanted to answer about the effectiveness and impact of MSU BEST. But how to get the data they needed?

Kevin: A challenge I see is getting the data we need to match interventions with outcomes. So far, we've thought about measuring the "dose" of MSU BEST career development interventions on trainees, partly in homage to Stephanie's work as a pharmacologist. We need to know what dose our trainees are receiving, so we need to figure out how to keep track of what trainees do during their BEST time and how that compares with the outcomes of the cross-campus Windrose surveys that the NIH requested. Those outcomes will help us see where our trainees land, but our data would help *us* understand what helped *them* in getting there. We also might learn more about what in each graduate program's culture and activities influences students.

Julie: We need an easy way for them to log their activities; the easier it is, the more likely it will happen. I doubt any of them are looking for more work to do or more things to keep track of.

Stephanie: And it has to keep track of not only the time they are involved in each, but also the type of activity and *when* they did it during their participation in the BEST program. Does something exist like this already?

Kevin: Not in the form we need it. There are many strategies people have used in other research studies, but I think we have an opportunity to build a tool that might work.

Julie: What specifically do we need it to do?

Stephanie: One, it has to be easy; it needs to allow trainees to input their own data. Two, it needs to be mobile; maybe an app or, more likely for cost reasons, a mobile-accessible, database-driven website that works well with phones.

Kevin: We can pre-populate the options so that we can control the data we need—have them select from specific activity types, for example, so we can restrict the possible data they can input. That will prevent a lot of misinterpretation and cleaning later on.

Julie: With a good design, we can make it easy for trainees to help us get the data we need for some great analysis later on.

Stephanie: And they're scientists who understand the need for good, clean data. As long as we make it easy for them, I think they will see the importance of their contributions.

Julie: We can't just make it "one more thing" they have to do, though.

Stephanie: Definitely not; there is real value in them tracking their own activities. They already need to report various activities if they're part of a training grant, and, for many students, their annual review includes this information.

A tool like this is good for us, yes, but it's also good for them: it can be a training diary of sorts.

And … I think I know just the right person to call.

(Stephanie picks up her phone and speaks into it)

Stephanie: Calling Ryan Doom at WebAscender!

(FADE TO GRAY, SPOTLIGHT ON NARRATOR)

Narrator: The BEST Action Inventory, or BAI, was created with WebAscender, and BESTies used this growing tool as a log or diary of their career development activities both within and outside of MSU BEST.

Starting at the very beginning of Year Two of the BEST grant, MSU BEST began plans to sustain BEST activities once the grant finished in September 2019. That meant engaging with many of the stakeholders described above and discussing how to proceed, what they need to change, and where they might look for ongoing support.

MSU BEST was housed in the Graduate School, welcomed there under Karen Klomparens, now Dean Emerita. Thereafter, it was supported by Interim Dean Judith Stoddart and then by Dean Thomas Jeitschko, who became the dean in Year Three of MSU BEST. All three of the involved deans agreed to sustain BEST during and following the grant period, and all agreed that locating BEST in the central Graduate School—versus a more discipline-specific college or unit—would be the ideal, especially as BEST became institutionalized. It would also allow for the continued support and collaboration from the Office of Ph.D. Career Services and Graduate Life and Wellness.

Through Years Three and Four (cohorts III and IV), they discussed how to make this program available to more students in the biomedical sciences at MSU, and potentially to disciplines outside of biomedical sciences. They reached a consensus that—to maintain the integrity of the experiment—they would keep BEST with a biomedical focus during the grant years. After, however, they would gradually test out and explore different options, like opening admissions to non-biomedical trainees, holding non-cohort rolling admissions, or decreasing the number of required externships/internships from two to one.

For the moment, however, they found that focusing their program on biomedical sciences was particularly beneficial for creating truly novel externships. These experiences have been born through the time and effort invested in building bridges with internship/externship partners, who are committed to keeping and expanding these experiences.

Their commitment is to take the most successful core elements of BEST and adapt them to different disciplines (e.g., social sciences).

(BLACKOUT)

(END OF SCENE)

Act II: Scene V

September 2018

SETTING	*Outside Chittenden Hall, home of the MSU Graduate School and Postdoctoral Office.*
AT RISE	*STEPHANIE stands alone.*

Stephanie: MSU BEST initiated its fifth and final year of the NIH-supported BEST grant in September 2018. This year, we started transitioning to some of the innovations we have planned to use for the institutionalized BEST, which we call BEST 2.0 (Fig. 4.7).

Instead of one large BEST cohort, we're bringing trainees into smaller groups who will experience each workshop, seminar, and guest speaker together. We orient them to BEST together and then support them as they go through the program at their own pace.

We have phased out the individual PI stipend; institutional funds will still come from the Offices of the Vice President for Research and Graduate Studies to support BEST, but they will directly fund student activities instead of offsetting costs for PIs or thanking them for their support. I am generally happy to report that removing the stipend has not diminished faculty support for MSU BEST. I think most faculty members have realized that the time commitment for their trainees is not onerous, and they have seen or heard their colleagues talk about how trainees are not missing that much time out of the lab. So, I am glad that we appear to have institutionalized the program without having these inducements in place. In fact, we've been able to collaborate with six other BEST institutions in

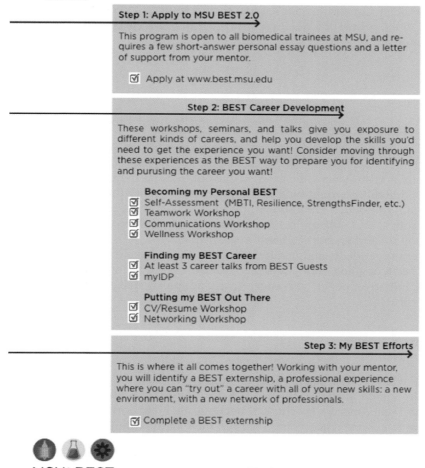

MSU BEST 2.0

MSU BEST 2.0 is a program to support biomedical trainees—Ph.D. students and post docs—in developing the skills and experiences to find, and succeed in, careers of their choosing. This program goes at your pace, and allows you to develop the path that serves YOUR interests.

Step 1: Apply to MSU BEST 2.0

This program is open to all biomedical trainees at MSU, and requires a few short-answer personal essay questions and a letter of support from your mentor.

☑ Apply at www.best.msu.edu

Step 2: BEST Career Development

These workshops, seminars, and talks give you exposure to different kinds of careers, and help you develop the skills you'd need to get the experience you want! Consider moving through these experiences as the BEST way to prepare you for identifying and purusing the career you want!

Becoming my Personal BEST
☑ Self-Assessment (MBTI, Resilience, StrengthsFinder, etc.)
☑ Teamwork Workshop
☑ Communications Workshop
☑ Wellness Workshop

Finding my BEST Career
☑ At least 3 career talks from BEST Guests
☑ myIDP

Putting my BEST Out There
☑ CV/Resume Workshop
☑ Networking Workshop

Step 3: My BEST Efforts

This is where it all comes together! Working with your mentor, you will identify a BEST externship, a professional experience where you can "try out" a career with all of your new skills: a new environment, with a new network of professionals.

☑ Complete a BEST externship

MSU ›BEST
2.0
Broadening Experiences in Scientific Training

Find out more at www.best.msu.edu!

FIG. 4.7   Flyer used for recruiting more widely to MSU BEST 2.0 beginning Fall of 2018. It outlines some of the modifications we intend to incorporate in the institutionalized version of BEST.

conducting a survey (created here at MSU by Kevin Ford, Rich Schwartz, Julie Rojewski, and Dia Chaterjee) that addresses these concerns. The results of this survey show that the biomedical faculty really are supportive of their trainees in their quest to find multiple types of biomedical careers [3].

We continue to capture data with BAI. We continue to actively reach out every year to new students, to speak at faculty meetings about the continuation of BEST, and to engage with new internship/externship partners.

We are immensely grateful to NIH for letting us run this experiment, and to the collaborators in the BEST Consortium for sharing their resources and challenges. Most of all, we are grateful to the people at MSU — especially the BESTies themselves — who joined in this grand experiment.

(Curtain)

Epilogue: Takeaways and helpful hints

Not everyone will engage, and that's ok

Not all faculty and students were as excited about BEST as the BEST team was. We underestimated how difficult it would be to share information about the program, let alone overcome resistance. We flooded email boxes, plastered walls with posters, talked with faculty and student groups, and still could have done more to connect with faculty and students, grab their attention, and encourage them to participate.

In our surveys of students and faculty, we know that we have not achieved 100% saturation. This will be an ongoing effort. Trying to reach everyone is an admirable goal, but, when starting a new initiative, begin with eager "early adopters" and close colleagues willing to support your efforts.

Listen to your naysayers

When you start a new initiative like BEST, you might encounter plenty of people who disagree with what you're attempting to accomplish and/or who disagree about the approach you're taking. Listen to them… but don't let them prevent you from moving forward. Listen to their concerns so you can decide if they are worth addressing and so you will know what to change when you go back to them in the future to try to get their buy-in.

For example, one piece of feedback both we and other BEST Consortium institutions heard was that there were insufficient data about faculty views on professional development programming like BEST. How could we design something that engaged faculty unless we knew what faculty members thought? That feedback did not prevent us from starting BEST, but it did motivate us to collectively survey faculty members and report about their attitudes.

Use existing resources to help you build up

MSU BEST was an experiment, one site among 17 being studied by the NIH through the BEST Consortium. We collected data and identified variables and treatment groups. We made data-driven decisions about how the program would be offered, how it would continue, and how what we saw aligned, or didn't, with our goals.

We were fortunate to have resources — programs already invented — that could serve as a foundation for MSU BEST. We did, however, have to develop many other things, such as the BAI, that can readily be used by others outside of MSU.

Like the other BEST Consortium members, we are eager to share our findings and resources (including BAI). If you aren't finding the data you need in our journal articles, reach out to us. The NIH BEST homepage[1] houses great resources from all 17 campuses and directs you to contacts at each program for more specific inquiries. The BEST Consortium website is itself a great resource to find ideas — ideas that have been tested for at least five years — so don't hesitate to use them.

In other words, don't reinvent the wheel: Use BEST Consortium resources (including MSU's BAI) and tailor them to your own culture and context: *ADAPT, don't just ADOPT.*

Tap Ph.D. alumni and their expertise

MSU alumni have been remarkably supportive when we have reach out to them. But the truth is, we continue to struggle to maintain close connections with alumni and to track where their careers and lives take them. Other BEST programs have been more aggressive than MSU in this regard, though, so you might want to refer to those institutions for more extensive help in this arena. Tools like LinkedIn make that a little easier, but MSU is like many universities; we only recently began engaging graduate alumni.

PIs are often the best source of information and connections to alumni. Do not hesitate to ask for their assistance and for introductions. Alumni can be a rich resource for career-related talks, internship hosts, and other networking.

We are learning that it's never too early to connect with alumni. Work with your alumni relations teams and engage with PIs who are particularly good at tracking their trainees. Your alumni have taken many paths, so follow up!

Be prepared for all types of trainees

Trainees with vastly different levels of motivation will come through your program. There will be those who are eager and those who are reluctant to start. There will be those who are there because their PI encouraged them to participate and those who found the program on their own, with or without PI support. There will be various levels of hand-holding needed and required by trainees who seek out these resources.

Make sure you have supports in place for students: sometimes the career exploration process can be stressful. Do not hesitate to connect with your student mental health resources for support or to refer students whose concerns about their career move beyond discussions about career options.

Engage widespread institutional support

Our BEST team engaged with stakeholders not only to build the program funded by the grant, but with an eye toward institutionalization. It made sense in an MSU context for MSU

---

[1]http://www.nihbest.org/.

BEST to be in the Graduate School, so engaging the support of the Dean of the Graduate School was key. Engaging with leaders across departments, colleges, and other units help build support and brought valuable diversity to the program. It meant we had a lot of good input from day one, such that we were prepared when it came time to make decisions about how to evolve the program and how to institutionalize it, as we had been collecting feedback all along from people who felt connected and committed to supporting it.

Don't wait until you are asking for money or other support to bring people into a new initiative. Include as many offices, with diverse perspectives, along the whole way. They will be able to follow along and be familiar — and invested — in your success.

Creating an internship/externship program for biomedical scientists is possible, even if a given campus is in an area not rich with private industry

Some cities like Boston, San Francisco, New York, and similar locations have rich research communities that may make partnerships more plentiful, but it is possible to develop something useful and meaningful in other areas.

You will, however, need to be flexible and creative: students committed to working in pharma may have to work in science-adjacent businesses or with your university's tech transfer office. That will expose them to business skills, even if they aren't explicitly in "big pharma." Encourage them to engage with their mentors and to identify research partnerships, colleagues, and friends who work in interesting fields, and to explore the possibility of designing something new, innovative, and flexible. Emphasize that the focus is on practicing new skills in new contexts, not simply creating a first-step toward a predetermined job in a specific field or company.

We are grateful to have had this opportunity. Thank you.

FINIS

Acknowledgments

MSU BEST was funded by the NIH, grant number DP7OD020320.

References

[1] Kindea Labs website. March 17, 2019. Available from: http://www.kindealabs.com/.
[2] MSU BEST website. March 17, 2019. Available from: http://best.msu.edu.
[3] Watts SW, Chatterjee, Rojewski JW, Reiss CS, Baas T, Gould KL, Brown AM, Chalkley R, Brandt P, Wefes I, Hyman L, Ford JK. Faculty perceptions and knowledge of career development of trainees in biomedical science: what do we (think we) know? PLoS One 2019;14(1). Available from: https://doi.org/10.1371/journal.pone.0210189.

New York University Science Training Enhancement Program

Arthee Jahangir[a], Keith Micoli[a], Christine Ponder[b], Carol Shoshkes Reiss[c]

[a]Office of Postdoctoral Affairs, New York University Grossman School of Medicine, New York, NY, United States; [b]Research Affairs, Postdoctoral Affairs, New York University, New York, NY, United States; [c]Department of Biology, New York University, New York, NY, United States

NYU STEP overview

The New York University Scientific Training Enhancement Program (NYU STEP) was designed as a three-phase effort to improve the career readiness of postdoctoral and graduate trainees. Our overall goal was to bridge the gap between perceived career opportunities and the actual careers most biomedical Ph.D.s ultimately pursue. NYU STEP engages trainees early to help them with actively planning their own careers, assessing their personal values, and translating those values into individual goals, all while introducing them to the diverse career opportunities that await them.

In the first phase, trainees participate in the career-training course, "Hope Is Not a Plan: Taking Charge of Your Career." This workshop requires trainees to create an individual development plan (IDP) for their careers and introduces the four broad career tracks we developed: for-profit industry, nonprofit and government, communications, and academia. The program continues in Phase 2 to help trainees develop transferable skills necessary for success within or beyond academia, such as time management, conflict management, communication skills, and professionalism. Phase 2 also provides more detailed, career-specific skills through a number of courses and seminars. Phase 3 supports all trainees through their job search and their transition to positions outside of NYU.

NYU STEP built upon successful programs developed at NYU but was a significant advancement in the formalization of career training and development of our trainees as science professionals. Starting with Phase 1, NYU STEP evaluates trainee awareness and knowledge of different careers at each program stage and transition in the program to

BEST: Implementing Career Development Activities for Biomedical Research Trainees
https://doi.org/10.1016/B978-0-12-820759-8.00005-X

determine the effectiveness of the program and to better track the career outcomes for all participants, including the traditionally hard-to-track postdoctoral scholars. NYU STEP encouraged trainees to push their careers forward more rapidly, and our intended outcomes included shorter time to Ph.D. and less overall time spent in postdoctoral training.

Background

The report of the NIH Biomedical Research Workforce Working Group, chaired by Shirley Tilghman, drew a clear conclusion that even among biomedical Ph.D.s, many did not attain the primary career goal that their training supposedly leads to — namely, becoming a tenured professor and a primary investigator at a research university [1]. The conclusion was not novel, as the National Research Council Committee came to the same conclusion in 1998 [2] when it was chaired by Tilghman herself, in addition to many others since that time [3–6]. A great deal of effort has been expended in the last 15 years to improve scientific training, and many noteworthy accomplishments have been achieved. In recognition of the difficulties that graduate schools encountered in the face of a changing employment landscape, the Association of American Medical Colleges formed the Graduate Research Education and Training (GREAT) group in 1996, which then formed the Postdoctorate Leaders Section in 2002. The formation of the National Postdoctoral Association (NPA) in 2002 was a watershed moment that pulled together key stakeholders in scientific training and gave them a vehicle to address problems collectively. Together, GREAT and the NPA have spearheaded the movement toward diversifying the training of biomedical scientists and have driven the formation of postdoctoral offices nationally.

Despite the efforts of these and other groups, such as the NIH and NSF, daunting challenges remain to be addressed if we are to fully realize the potential of the US scientific enterprise. Academic hiring trends display a marked shift away from traditional tenured positions in basic research toward more clinical and non-tenure track faculty, and global increases in the production of Ph.D.s have placed ever-increasing stress on the workforce pipeline. The economic recession and resultant collapse of private sector careers for Ph.D.-level scientists severely limited the ability of this sector to absorb the highly trained workers produced by academia and forced many of them to consider careers that they may not have known existed when they began their Ph.D. studies. The job market realities of today are very different than they were even a decade ago, and they are unrecognizable to past generations of scientists.

Barring a complete overhaul of the Ph.D. system (not only in the US but globally), the best way to address this new reality is to better train biomedical Ph.D.s for the realities of the job market. As the Tilghman report highlights, actual unemployment among Ph.D.s is very low. Ph.D.s are finding jobs and — more than that — rewarding careers; however, the NYU STEP leadership believes trainees are finding these jobs and careers later than they should. Increasing the awareness of the breadth of careers available to Ph.D.s may motivate students to finish faster and move straight into careers without a postdoctoral experience, allowing them to make a greater economic contribution earlier in their lives.

NYU STEP proposed a new training model for biomedical doctoral students and postdoctoral fellows based on several years of experience developing innovative career development resources. Trainees from both NYU campuses, Langone Health (LH) and Washington Square (WS), participated. Our overall goal was to prepare our trainees for professional positions after their training was completed, while recognizing the scarcity of tenure-track faculty positions to which many aspire.

We established an open but comprehensive training program that moved trainees through three phases and into four distinct − yet broad − career tracks. An additional intended benefit was to disseminate successful elements of the program, providing benefits to trainees far beyond NYU. We approached our efforts scientifically and systematically, collecting data on trainee background, interests, participation, satisfaction, and outcomes. This assessment of our trainee population allowed us to evaluate and modify NYU STEP based on concrete data and will allow objective comparisons to other programs.

We proposed creating and implementing a comprehensive training infrastructure and a defined career planning and exploration pathway to lead to a more efficient, effective, and satisfying training experience. We helped trainees identify career goals early on, provided resources to actively pursue specific career options, and established an outcomes-tracking system for all trainees. This model of scientific training is broadly applicable across different institutions and creates a hub of innovation that serves the entire biomedical education and training system.

Our three-phase career development program is described briefly below:

Phase 1. Development and implementation of the 10-week IDP Course for graduate students and postdocs, offered biannually.

Phase 2. Development of general and career-specific skill-building courses and workshops:

  a) Skills in communication, negotiation, and professionalism, including leadership and time and conflict management.

  b) Track-specific skills for careers in for-profit industry, nonprofit and government, communications, and academia.

Phase 3. Preparation of trainees for career transition into either academic or non-academic careers.

We strongly believe that this training model could be replicated at almost any research institution with significantly large populations of scientists in training and an administration committed to implementing this model (Fig. 5.1).

Phase 1: Career planning and exploration

Career planning

Trainees go through a process of self-evaluation and exploration of the career options that are available to Ph.D.s. They also participate in a formal career-planning course using the web-based myIDP program [7].

NYU STEP's IDP Course focuses on four defined career tracks:

  1. Careers with for-profit companies
  2. Careers with government and nonprofit organizations
  3. Careers in science communication
  4. Careers in academia (broadly defined to include teaching and administration)

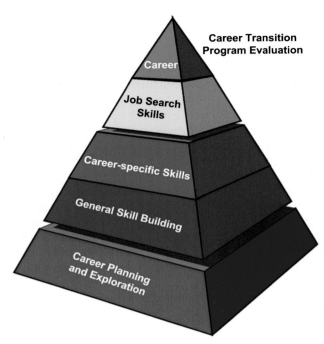

FIG. 5.1 Schematic illustration of the NYU STEP phases. The illustration above depicts the sequence in which participants move through the program. Beginning at the base of the pyramid, **Phase 1** (red) includes the Career Planning Course and Career Exploration Panel Discussions. **Phase 2** (blue) includes the development of career-specific skills and general skills, such as communication, negotiation, leadership, and professionalism through coursework (e.g., "Fundamentals of Teaching," "Careers in Consulting," "Preparing for Industry," and more). **Phase 3** (yellow and green) teaches resume/CV writing, interview skills, and networking. Phase 3 also includes company visit days and informational interview days, and ends with program evaluation of career outcomes and follow up with alumni to serve as career mentors and panel discussants for future Phase 1 events.

Although the IDP Course was originally 1.5 hours per week for 10 weeks and was taught to a mix of graduate students and postdocs, we later shortened it to eight weeks and created separate sections for graduate students and postdocs after initial feedback. The student curriculum was modified to emphasize planning for graduation and next steps and to deemphasize resume writing and interview preparation.

A 2016 survey of trainees who had taken the IDP Course in 2013—15 had a 61% response rate (94 respondents out of 154 who took the course). Of these, 32 (34%) were in different positions two to three years after taking the course; 20 respondents (21%) reported that the IDP Course was helpful in their transition to a new position. Only 4 respondents reported that the course was not helpful, while 5 respondents were unsure. Overall, 57 trainees (60%) responded that the IDP Course helped clarify their career goals compared to 23 (24%) who said it did not help.

These preliminary results are very positive, and we are excited to continue to track participants over a longer time frame and gather a much larger sample size.

Career path exploration

After the career-planning course has allowed trainees to explore their options, they have the opportunity to further delve into career exploration by participating in career-focused clubs, like consulting, biotech, and programming clubs, attending career panel discussions and networking events, and taking additional courses. These additional courses include NYU STEP courses listed in the section below, but also courses offered by other NYU departments or by regional resources such as the New York Academy of Sciences.

Phase 2: <u>General skill building and development of career-specific skills</u>

Once trainees define broad career areas of interest and identify potential career targets, they have the opportunity to develop additional skills that go beyond the technical skills they develop through their research training. They develop **general and career-specific skills**, including communication, professionalism, networking, negotiation and creativity, by participating in formal courses to help in their goal of obtaining their next position. Many of the available courses were developed in partnership with employers from each track to ensure that their contents were appropriately tailored to best prepare trainees to become competitive job applicants.

Providing these courses ended up being the most important effort of NYU STEP. Career exploration is important, but we found many Ph.D. students had already done some of it themselves and already had an idea about where they might want to go. However, many of them felt unprepared for their desired careers, especially if those were outside of academia; providing skill development courses that increased knowledge and confidence in their chosen career areas was thus a key success of the program, based on qualitative responses to post-course surveys.

Participants begin developing in-depth, career path-specific skills in semester long courses. We offer the following courses, separated by career track:

For-profit careers

- The Business of Science
- Biotechnology Industry: Structure and Strategy
- Drug Development I and II (taught by faculty at the NYULH Clinical & Translational Science Institute)

Nonprofit and government

- Science Policy
- Science Diplomacy

Communications

- Practical Skills in Medical Writing
- Science Communication I and II
- Confident Communicators of Science

Academia

- Fundamentals of Teaching
- Laboratory Management
- Grant Writing

Summary of NYU STEP programming and participation

Figs. 5.2 and 5.3 summarize the growth of NYU STEP, and demonstrate the very wide participation across the program. One initial reservation we had was whether the proliferation of courses and workshops would prove overwhelming to trainees, and result in dwindling participation rates. We see in Fig. 5.2 that this was not the case, although we did see a plateau from Years 2 to 4 as the programming changed from low-commitment activities (workshops and career panels) to higher-commitment activities (half- to full-semester courses). By Year 5 of the program, we directly organized or co-organized 25 courses — more than double the number of courses that were available during the program's first year — each requiring 10–40 hours of contact time. Registration increased from 525 participants in Year 1 to 1,653 in Year 5, which reflects a greater-than-3-fold increase. The fact that our registration exceeded the actual number of trainees in our combined programs (about 1,000) was largely due to the fact that many trainees enrolled in multiple courses and to the fact that we opened up our courses to non-NYU participants from across the New York City area. Non-NYU participants were recruited from partner schools through email and listserve advertising and by emailing previous participants of non-STEP events, such as the biennial What Can You Be with a Ph.D.?, the largest career symposium in the country.

Fig. 5.3 summarizes the registration data for the 22 courses offered during the 2017–18 academic year, and shows that of 613 total registrants, 514 attended at least 1 session of the

FIG. 5.2 Broadening training opportunities. Over the past five years, NYU STEP organized or co-organized 186 events. The numbers in the center of the circles indicate the total number of events per year, broken down by event types (indicated by color). In addition to those, we also co-hosted career development events with our partners from the Wasserman Center for Career Development, student/postdoc clubs, the Biomedical Entrepreneurship Program, the Sackler Institute of Graduate Biomedical Sciences, the What Can You Be with a Ph.D.? symposium, and the New York Academy of Sciences.

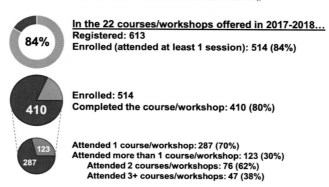

In the 22 courses/workshops offered in 2017-2018...
Registered: 613
Enrolled (attended at least 1 session): 514 (84%)

Enrolled: 514
Completed the course/workshop: 410 (80%)

Attended 1 course/workshop: 287 (70%)
Attended more than 1 course/workshop: 123 (30%)
    Attended 2 courses/workshops: 76 (62%)
    Attended 3+ courses/workshops: 47 (38%)

FIG. 5.3 Summary of the registration data for the 22 NYU STEP courses and workshops offered during the 2017–18 academic year. Of 613 total registrants, 84% (514) completed the course or workshop they enrolled in, an increase from 70% during the 2016–17 academic year (data not shown). Of the 514 registrants, 410 were individual attendees. Seventy percent of the individual attendees (287) attended 1 course or workshop, and 30% attended more than 1 course or workshop (123).

course and were considered "enrolled." Based on the courses where we had accurate registration and attendance data, this dropout rate was consistent across all courses. The completion rate improved year after year, approaching 80% in 2017–18 (Fig. 5.3). A key takeaway, however, is that although more people registered for a course than actually engaged with it, those who did engage were highly likely to complete it.

Phase 3: Career transition and outcomes tracking

Career transition

In the final phase of program participation, we help trainees begin their job search, building from the work done in the IDP Course. A major source of support has come from the NYU Wasserman Center for Career Development, which provides one-on-one resume reviews and hosts employer networking events, resume-writing workshops, and other professional skill-building events.

Trainee-led interest groups and clubs have also proven highly successful and sustainable. NYU-STEP provides funding and logistical support to these organizations and communicates opportunities to the broader trainee population to encourage participation. These groups play an important role as a peer-to-peer mechanism of practicing the specific skills needed to transition to a job outside of academia. The Consulting Club, for example, has evolved into a university-wide group with over 100 members and has begun collecting dues to support its activities. The club's main event is a case competition that is organized in conjunction with consulting companies and is open to all NYC graduate students and postdocs. Another example is the Biotech Club, which hosts networking events periodically, provides resume and interview practice sessions, and organizes an NYC-wide biotech expo each year.

The Medical Communications Course has proven highly successful in enabling trainee transition to medical writing careers, largely thanks to realistic writing tests that are evaluated by working professionals in the field. By 2018, 20 individuals who had taken the course had either graduated or completed their postdoctoral fellowship, including 12 (60%) who held positions in the medical communications field. Even with the understanding that this group self-selected, it is very encouraging to see these results, and we are expecting that number to increase as we modify the course to be more rigorous. It should be noted that 2018 marked the first time we opened this course to non-NYU participants, and the demand for the class surpassed the established cap for both the 2018 and 2019 classes, a reflection of the substantial demand for this training.

Outcomes tracking

Phase 3 of NYU STEP involves helping trainees transition into their careers and tracking where they go. To support trainees with job search help, NYU STEP provides career counseling services that cover topics such as interview preparation, CV or resume writing, cover letter writing, and networking to help identify potential positions. To help with outcomes tracking, we encourage trainees to maintain contact with NYU STEP using social media, including LinkedIn, Facebook, and Twitter, once they leave NYU. We also conduct periodic surveys of NYU STEP graduates to obtain job placement and career advancement information.

In addition, we continue to perform longitudinal surveys of our alumni, since it may be difficult to judge the true value and effectiveness of the program until participants have more experience in the broader job market.

Pre-existing outcomes data

NYULH's Sackler Institute of Graduate Biomedical Sciences has been a leader in fostering early career development for graduate students and postdocs and was one of the first institutions to create a formal office focused on postdoctoral training in 1996. The corresponding Office of Postdoctoral Affairs at NYUWS was established in 2011 to serve the rest of the NYU campus, which houses postdocs in Biology, Chemistry, Neural Science, and Psychology; NYU STEP serves these postdocs as well.

These two offices have surveyed their postdoc populations to ask about career expectations. The results show that a significant number of postdocs from both campuses are unsure of what they want from their career. When asked about their career goals, a large percentage (19% at NYUWS and 22% at NYULH) answered "not sure." In addition, many postdocs (58% at NYUWS and 42% at NYULH) identify "tenure-track faculty" as their goal, but the data do not support the idea that all of these postdocs will eventually obtain faculty positions.

NYU STEP addresses the significant portion of the population who have no idea what they'll do next as well as those that have failed to imagine other possibilities, helping them become more aware of other opportunities should their first career choice turn out to be unattainable. It is noteworthy that 20%−30% of the postdoctoral population does have specific career goals that are not becoming a PI. NYU STEP provides them with essential support, training, and networking opportunities that otherwise might not be available in an academic setting.

The Sackler Institute of Graduate Biomedical Sciences tracked all 285 of its Ph.D. alumni from the decade prior to the start of NYU STEP (2002–12) and identified their first position after graduate school. The majority of graduates (72%) went on to a postdoctoral fellowship as their first position. For those who completed training (finished their postdoc or additional education), 55% worked in the for-profit sector, including the pharmaceutical/biotech industry, management consulting, intellectual property law, investment banking, or other. Government and non-profit organizations employed 11% of our post-training graduates, and the medical communications sector employed 10%. Broadly, academia employed 22% of Sackler graduates; 5% were in tenured or tenure-track faculty positions and 17% were in non-tenure track faculty, administration, or research scientist positions. Only 2% of graduates from 2002 to 12 reported being unemployed, and all of those individuals were either in transition from their Ph.D. laboratory or were new parents who chose to take time out of the workforce to care for their families. These data are very much in alignment with national trends, which indicate that 70% of biomedical Ph.D.s pursue a postdoctoral position, 18% pursue industry research, 6% pursue government research, and only about 2% remain unemployed [8].

NYU WILD

Another exciting and highly impactful aspect of NYU STEP was developed with the input of trainees themselves, who found that leadership development opportunities were rather limited, particularly for women. To address this need, we developed a Women's Intensive Leadership Development (WILD) program designed to provide women graduate students and postdocs with practical leadership skills in an outdoor setting, and although it requires tremendous effort and resources, it appears to be one of our most impactful programs.

NYU STEP launched NYU WILD in 2017. The mission of the program is to achieve diversity, equity, and inclusion in the scientific workforce by empowering early career women scientists with self-confidence, leadership, and team-building skills. Participants (6–10 per cohort) spend seven months developing a strong sense of their leadership abilities by working with other aspiring leaders and embarking on an eight-day, 100-mile thru-hike in the Adirondack backcountry.

NYU WILD comprises three parts: leadership workshops, practice hikes, and the final thru-hike. Participants are required to attend monthly workshops that cover a range of leadership topics such as trust-building, unconscious bias, emotional intelligence, conflict management, problem-solving in a team, communication, and advocacy. The workshops consist of both didactic and activity-based learning. Wilderness principles and survival skills are also covered during the workshops, including "Leave No Trace" principles, first aid, risk management, and expedition behavior.

Every month, the participants hike in local state parks to develop their hiking and camping skills. The participants learn how to mentally and physically endure the trails while building trust and forging bonds with the other team members. Increased cognitive, psychological, and physical skills are required for training for a thru-hike trip, which explains the necessary length of the WILD program. The program's length enables team members to build strong relationships in a cooperative, non-competitive environment.

For the final thru-hike, each member assumes a functional role — logistics, safety, fitness, communication, or leadership — which then guides them into becoming a cohesive team. By the time they take the 100-mile hiking trip, the participants are able to carry backpacks containing up to 40 pounds of essentials (including food) and hike between 8 and 19 miles a day. Each day, the team hikes about eight hours and engages in different leadership discussions and activities after setting up camp.

The NYU WILD program has been met with great enthusiasm and success. WILD alumnae begin this challenging journey because they have personal reasons for wanting to develop their leadership skills, but the challenges of being a woman in STEM and dealing with discrimination and societal shortcomings of providing women scientists with equal opportunities are common themes. WILD creates a safe space for women to share their experiences and support each other. The participants can showcase their strengths in a team as they embark on a literally and figuratively long journey.

The WILD program has had 10 participants over the past two years, and all 10 have successfully completed the 100-mile trip. WILD expanded in 2019 to include two new groups: WILD Lite and TRAIL. WILD Lite is a program in which women scientists complete a 40-mile trek over four days. TRAIL, Training Resilient and Inclusive Leaders, is a program for scientists of any gender or non-gender to complete a 60-mile trek over four days. These programs follow the same curriculum and practice structure as WILD.

As the program grows, we will conduct an evaluation of the curriculum and outcomes of the participants. We are collecting feedback to design workshops for optimal knowledge building and application. The programs' desired outcome is for each participant to understand how to be an effective leader, team player, and supportive scientist who will work to achieve diversity, equity, and inclusion. Finally, participants should also be able to communicate why their scientific contributions are important to society and nature.

Although the sample size is currently small, the results are quite impressive. The 2017 WILD group had six participants, including Arthee Jahangir, Assistant Director of Postdoctoral Affairs at NYU Langone Health. Four of the six participants have successfully transitioned to independent careers, including two tenure-track faculty positions, a medical writer, and a bioinformatician. Jahangir went on to further her leadership skill development as part of the 2019 Homeward Bound program [9], which brings together 80 female scientists from around the world on a 21-day expedition to Antarctica.

All four of the 2018 WILD participants are all still training at NYU but have all shown tremendous growth in their leadership skills. They collectively created an NYU-wide Hiking Club, which organizes monthly hikes and has over 100 members. They are also co-leading the 2019 WILD Lite and TRAIL trips, which has the added benefit of demonstrating to new participants the level of expertise that one can develop by going through the program.

Closing thoughts

NYU WILD and all of the other programs and resources made possible by the NIH BEST grant greatly enhanced our ability to begin to change the academic training culture at NYU and align it with the current reality. We more than tripled participation in our offerings over the course of five years, fostered cross-discipline and cross-campus collaborations, and have begun to develop a regional hub for career development that maximizes efficiency and effectiveness. In the past, one of the great barriers to progress at NYU has been the tendency of each school to exclusively focus its efforts and attention internally, which limits their ability to learn from each other and to leverage their successes. We feel that NYU STEP provides a model for partnering that is cost effective, sustainable, and trainee driven. These are all important parts of the puzzle, and although we have not completed our aims, we have made a solid start.

Acknowledgments

The authors would like to acknowledge the many contributions and collaborations across New York University that made this work possible. This includes Dafna Bar-Sagi, Senior Vice President and Vice Dean for Science at NYU School of Medicine, and Michael Purugganan, NYU Dean for Science, both of whom provided full support and additional resources to ensure NYU STEP's success. The NYU Wasserman Center for Career Development was an instrumental partner in providing career transition assistance to students and postdocs, as well as connections to external employers. The NYU Clinical and Translational Science Institute provided access to courses and other resources that were extremely popular and helpful to our trainees. Finally, the staff of the Office of Postdoctoral Affairs provided essential administrative support, without which none of our programs could have been possible. The authors wish to acknowledge their contribution with great thanks. This work was supported by NIH grant DP7 OD018419.

References

[1] NIH Biomedical Research Workforce Working Group. Final report. June 14, 2012. Available from: http://acd.od.nih.gov/Biomedical_research_wgreport.pdf.

[2] National Research Council. Trends in the early careers of life scientists. Washington, DC: The National Academies Press; 1998.

[3] National Research Council. Enhancing the postdoctoral experience for scientists and engineers: A guide for postdoctoral scholars, advisers, institutions, funding organizations, and disciplinary societies. Washington, DC: The National Academies Press; 2000.

[4] National Research Council. Addressing the nation's changing needs for biomedical and behavioral scientists. Washington, DC: The National Academies Press; 2000.

[5] National Research Council. Bridges to independence: Fostering the independence of new investigators in biomedical research. Washington, DC: The National Academies Press; 2005.

[6] National Research Council. Rising above the gathering storm: Energizing and employing America for a brighter economic future. Washington, DC: The National Academies Press; 2007.

[7] Hobin JA, Fuhrmann CN, Lindstaedt B, Clifford PS, September. You need a game plan. Science Careers, 2012. Available from: www.sciencemag.org/careers/2012/09/you-need-game-plan.

[8] Sinche M. Next gen Ph.D.: A guide to career paths in science. Harvard University Press; 2016.

[9] About Homeward Bound. Available from: https://www.homewardboundprojects.com.au/about/.

Leadership and management for scientists

Tracey Baas, Steve Dewhurst, Sarah Peyre

University of Rochester, Rochester, NY, United States

Introduction

Moving from team member to leader is described as a transformation that results in a fundamental change in identity and point of view, which is often associated with stress and intense emotion [1,2]. Preparation and support are critical to the ability to successfully navigate this transition — yet are often underdeveloped.

Academic faculty members and industry group leaders are typically hired and promoted on the success of their research, publications, and funding, even though they will devote much of their time to teaching, mentoring and managing a laboratory group — tasks for which they have not necessarily been prepared and for which they may not possess a natural talent [3]. Most everyone acknowledges that research skills are very different from the skills needed for managing a lab, but new faculty are still expected to learn as they go, often modeling their leadership and management styles based on experiences with their own mentors regardless of the quality of those past mentor-mentee relationships.

In a survey of 3,200 scientists, the model of faculty learning as they go (i.e., picking up leadership and management skills on the job) was found to be ineffective and inadequate [4]. Lab members wanted more principal investigators (PIs) to take training courses in mentoring and management. Interestingly, PIs responded in the same way and wanted more support for mentoring and management. This makes sense because leadership quality is a strong predictor of group productivity and turnover and because the level of scientific integrity within any given lab environment depends largely on the leadership and management practices of the lead investigators [5]. Activities such as cultivating a lab culture, mentoring trainees, assembling teams, training and supervising staff, solving technical problems and improving work processes are all integral to good research [5–8]. Furthermore, individual career advancement is strongly influenced by leadership training — as evidenced by the fact that assistant professors at the University of California, San Diego, who completed a leadership curriculum

BEST: Implementing Career Development Activities for Biomedical Research Trainees
https://doi.org/10.1016/B978-0-12-820759-8.00006-1

were significantly more likely to be promoted while on the tenure track [2,9]. It is therefore disheartening that many academic institutions have been hesitant to make leadership training mandatory for their faculty, because of the perceptions that other "institutions don't rate [training] that highly" [4].

Intentional leadership and management practices should be introduced at an early stage of researchers' careers, and these practices should be continually refined along the way. Such instruction could be particularly beneficial at the graduate and postdoctoral stages in helping to prepare future faculty members as well as those pursuing non-academic careers. Indeed, doctoral fellowships and training grants from the National Institutes of Health (NIH) already require such training [10]. Ciampa and colleagues [2] observed that graduate students and postdoctoral associates have more time to reflect on key concepts and develop core skills before it becomes necessary to apply them. As a result, well-executed leadership development programs would be beneficial for all students and fellows, not just those on NIH training grants. Although some people argue that intense educational curricula leave little time for formal leadership and training experiences, it could also be argued that trainees have a more flexible schedule for professional career development than faculty members who are juggling more duties and responsibilities [5].

Although very little is known about leadership and management in scientific research, it is clear that leadership is particularly critical when the nature of work is complex and creative [5,11]. Across disciplines and career tracks, scientific leaders all engage in a complex array of behaviors to varying degrees. Some are more traditionally associated with management activities, such as planning, obtaining resources, providing feedback; and others are more traditionally associated with leadership activities, such as coaching, modeling, building relationships and providing socio-emotional support [5]. A major issue moving forward is defining the nature of training and development efforts for scientists.

Emphasis and a call to action

To address this lack of leadership training across the sciences, intentional leadership and management practices can be introduced during graduate and postdoctoral training. Traditional biomedical and STEM research training programs fail to adequately prepare trainees for the full diversity of careers that they will ultimately experience in the scientific workforce. To address that issue, NIH created a new program in 2013, titled the "NIH Director's Broadening Experiences in Scientific Training (BEST)." The University of Rochester (UR) was one of 17 institutes to receive funding under this umbrella, with the mandate of conducting "a series of experiments to identify new and innovative approaches to broaden career and professional development for graduate and postdoctoral training."

Our URBEST program was designed to provide a flexible, autonomy-supportive learning structure that allows pre- and postdoctoral trainees to better prepare themselves for diverse career paths by exploring different scientist personas and accumulating diverse skill sets. One aspect of career development that URBEST deemed applicable to all biomedical and STEM trainees, no matter what career they chose, was leadership and management. Because trainees already undertake a heavy academic course curriculum and spend a large number of hours working in their research laboratories, we understood that leadership and management training had to be presented within the context of current scientific endeavors and future careers, and developed a course designed to meet those needs: *Leadership and Management for Scientists*. We wanted students and postdocs to find immediate value in lessons learned so that they could test and apply some of their learned skills with UR colleagues, and help create interest and buy-in for the leadership development course with their peers (potential course registrants).

Our URBEST program was in a unique position to unlock institutional resources that have not traditionally served as venues for the education of trainees. Specifically, UR supports a broad range of non-academic units that meet essential institutional needs (e.g., technology transfer and clinical research offices, regulatory oversight committees, a cGMP manufacturing facility, etc.) that had not previously provided experiential learning opportunities and educational content. This "unlocking" approach was achieved with minimal incremental cost, and can be replicated easily at other institutions since analogous non-academic units are found at essentially all US research universities.

Another resource traditionally "hidden in the open" from graduate students and postdocs is senior leadership. Students and fellows may interact with some of these senior leaders in their capacity as scientists in research laboratories, but they typically do not hear about the roles these senior leaders play in supporting and advancing the university mission. The UR already taps these leaders to help mentor newly appointed faculty and departmental chairs, so we elected to reveal these role models and their leadership skills early in the career development of biomedical and STEM graduate students and postdocs. Again, this "revealing" approach was achieved with minimal incremental cost and can also be easily replicated at other institutions.

We anticipated interactions with trainees who had different learning requirements and time availabilities, so we wanted to provide a range of programming that would fit the needs of each individual learner. To do this, we developed and implemented three different forms of education programs — intensive, distributive, and on demand — to guide students and fellows toward becoming proficient managers and leaders, helping them to discover and build leadership qualities and interpersonal skills that work for them.

Intensive programming: Leadership and management for scientists course

Because we wanted our first training module to be a capstone in trainees' career and professional development, we partnered with the Center for Professional Development (CPD) in the UR School of Medicine and Dentistry. When we first started thinking about what we wanted the training to look like, we fell back on the traditional method of instruction: the semester-long course. Together, URBEST and CPD created a pass/fail, 15-week, interdepartmental course, IND493 *Leadership and Management for Scientists*, which is open to all *Ph.D.* graduate students, MS students, and postdoctoral associates in biomedical and STEM subjects at the UR.

As a first step, we discussed possible topics for the course with a range of colleagues. Several specific topics surfaced from these conversations: managing people, interviewing, resolving conflict, assessing personalities, and overseeing resources. Knowing that one individual would not be capable of teaching all the various components and skills associated with being an effective leader or manager, we then identified influential leaders or up-and-coming leaders at UR who would be qualified to address these issues and related subjects. They each accepted the invitation to teach, and selected one of the specified leadership topics or suggested their own based on their respective areas of expertise, such as cultural humility, personal mission, teamwork, handling change effectively, and more. In fact, many of these topics were areas of expertise that our instructors were commonly requested to speak about and already had slide decks for.

We were quite ambitious our first year and invited 18 instructors to give 23 classroom sessions (Table 6.1). The course was facilitated by two co-directors, a CPD representative and a URBEST representative (Tracey Baas, *Ph.D.*), who introduced the speaker of the day and who framed each session's topic in the context of the bigger picture of leadership and management.

As a whole, *Leadership and Management for Scientists* focused on human-centered strategies for effectively leading teams in biomedical and STEM academic environments, which could be put into use for any research-related career. We met for two 1.5-hour sessions each week for 15 weeks, during which the instructors incorporated multiple teaching styles and tools, including short videos, case examples, group work, traditional but truncated lectures, and lots of dialogue to investigate the chosen topics. Through a series of interactive activities, reflective writing, self-assessment instruments, and group discussions, trainees developed a repertoire of techniques for addressing issues that commonly arise within and between research groups. The activities helped to promote awareness of the participants' own styles of leadership and offered new approaches to explore.

To refine and improve our course offerings, students were assigned evaluations for each of the instructors and for the full course. All evaluations had to be completed in order to receive a passing grade. One of the most obvious glitches of Year One was that we scheduled too many assignments and class sessions. This was reflected in some of the initial course evaluations, as exemplified by this comment:

> Okay, there were WAY too many assignments. I appreciate that you need to have something to "show" for the class but I've taken three-credit courses before and had to do less work than this class! My recommendation is that instead of having to complete so many assignments, count the group project, individual projects and leadership assessments for grades. Additionally, award points for attendance. I think you lost a lot of people because they knew they could come to whatever they wanted without consequence. I think part of being a good scientist is time management and if you signed up for the course, you should make time for it, maybe two absences or something like that since people do go to conferences, etc.

Despite concerns about over-scheduling, some of the evaluative comments also noted areas that students and fellows felt were missing from the course content. Conflict resolution was a particularly important area, as noted in this remark:

> I would have liked to see more about conflict solving. It's important to understand your personality and how it fits into a team as well as how and why others may be reacting the way that they are. However, what do you actually do when people are butting heads? If you're part of a team and the leader isn't being a good listener and isn't holding people accountable, how do you approach those types of situations?

TABLE 6.1 Changes in the syllabus of the Leadership and Management for Scientists course through the first four years of URBEST.

Year One	Year Two	Year Three	Year Four	Year Four Personnel
Managing People[a,b]	Managing People[a,b]	Managing People[a,b]	Managing People[a,b]	Managing Vice Dean for Research (School of Medicine and Dentistry)[a,b]
Mentoring[b]	N/A	N/A	N/A	N/A
Cultural Humility	Cultural Humility	Cultural Humility	Cultural Humility	Organizational Development Specialist (Human Resources)
Personality Types Part I	Personality Types Part I	Personality Types Part I	Personality Types Part I	Director of Career Services (Graduate Education)
Personality Types Part II	Personality Types Part II	Personality Types Part II	Personality Types Part II	Director of Career Services (Graduate Education)
Emotional Intelligence Part I	N/A	N/A	N/A	N/A
Emotional Intelligence Part II	N/A	N/A	N/A	N/A
Mentors and Mentees in the Digital Age	Mentors and Mentees in the Digital Age	Mentors and Mentees in the Digital Age	Mentors and Mentees in the Digital Age	Executive Director of URBEST
Science Communication and Outreach[b]	N/A	N/A	N/A	N/A
Networking – How to NOT be Awkward	Networking – How to NOT be awkward	Networking – How to NOT be awkward	Professional Social Skills	Executive Director of URBEST
Entrepreneurial Mindset Part I	N/A	N/A	N/A	N/A
Entrepreneurial Mindset Part II	N/A	N/A	N/A	N/A
SBIR/STTR	N/A	N/A	N/A	N/A
Interviewing: Selecting Teams[a,b]	Interviewing: Selecting Teams[a,b]	Interviewing: Selecting Teams[a,b]	Interviewing: Selecting Teams[a,b]	Chair of Public Health Sciences[a,b]

(Continued)

TABLE 6.1 Changes in the syllabus of the Leadership and Management for Scientists course through the first four years of URBEST.—cont'd

Year One	Year Two	Year Three	Year Four	Year Four Personnel
Communicating and Resolving Conflict as a Leader[b]	Communicating and Resolving Conflict as a Leader[b]	Communicating and Resolving Conflict as a Leader[b]	Communicating and Resolving Conflict as a Leader[b]	Senior Associate Dean of Academic Affairs[b]
Personal Mission	Personal Mission Part I	Personal Mission Part I	Personal Mission Part I	Instructor for Technical Entrepreneurship and Management (Simon Business School)
Personal Mission/Innovation Creed	Personal Mission Part II	Personal Mission Part II	Personal Mission Part II	Instructor for Technical Entrepreneurship and Management (Simon Business School)
Innovation Creed	N/A	N/A	N/A	N/A
Overseeing Resources Effectively	Overseeing Resources Effectively	Overseeing Resources Effectively Part I	Overseeing Resources Effectively Part I	Research Administrator in Academic Affairs Office (School of Medicine and Dentistry)
Strategic Planning	Strategic Planning	Overseeing Resources Effectively Part II	Overseeing Resources Effectively Part II	Research Administrator in Academic Affairs Office (School of Medicine and Dentistry)
Teamwork and Collaboration[a,b]	Teamwork and Collaboration[b]	Teamwork and Collaboration[b]	Teamwork and Collaboration[b]	Associate Dean for Innovative Education (School of Medicine and Dentistry)[b]
Project Management[a,b]	Project Management[a,b]	Project Management[a,b]	Project Management with Teams[a,b]	Vice Provost Health Sciences Center for Computational Innovation (School of Arts, Sciences and Engineering)[a,b]
Handling Change Effectively	Handling Change Effectively	Handling Change Effectively[a,b]	How to Thrive In An Era of Digital Publishing[a,b]	Senior Associate Dean of Clinical Research (School of Medicine and Dentistry)[a,b]

[a]Course instructor is the PI of a research laboratory
[b]Course instructor is senior leadership

the T3 session, was first presented in 2016, received strongly favorable feedback from students, and was presented again in 2018. This approach was achieved with minimal incremental time and expense, and can also be replicated easily at other institutions, since students, fellows, staff, and faculty undertake analogous career development yearly at essentially all US research universities.

Other broad asks were (1) *"Do you know anyone to help our trainees with science communication and writing more manuscripts?"* (2) *"Would you be able to explain how to create an Individual Development Plan (IDP)?"* and (3) *"Could you lead any type of hands-on activity that would help our trainees?"*

The first ask was for a T32 Science Communication Supplement for one of our *Ph.D.* programs. Drawing on Baas's past experience as a science writer and editor and on her passion for the written word, we created a slide deck, *Writing Tune-up for Scientists*, relatively easily. The audience found the *Writing Tune-up* session extremely useful and requested a longer session and additional units.

The second ask was from another *Ph.D.* program and was relatively easy to address because URBEST already leads an IDP presentation. This session is coupled with a team of professors who come in to talk to students and who offer one-to-two (i.e., one professor to two students) advice on reaching IDP goals. The triad setup takes a little more planning, but the inclusion of two students has the advantage of making the interaction feel less like less a formal interview. By using a one-to-two ratio and five-minute conversations with three different triads — speed mentoring — this workshop has been a consistent hit.

The third ask was simply to improve the research seminar in one of UR's basic science departments, with no further specificity. In this case, we decided to implement URBEST trainees' favorite activity. The opening of the session starts by reenacting Tom Wujec's Marshmallow Challenge [15] and his ideas of collaboration, teamwork, and prototyping. We enhanced the session by including a meaningful message for *Ph.D.* students: that successful scientists use prototyping both in their laboratory with their experiments and in their lives with their careers. When we received evaluations like, *"Wow, that was much more useful than I had guessed,"* and, *"We need you to run more of these sessions,"* we knew that the session was a success.

An important added value of these on-demand sessions is that they motivate trainees to enroll in URBEST, to sign up for a course, or to just set up an appointment with URBEST leaders for a chat when they reach certain milestones or challenges in their training and lives.

It has also become clear that not all trainees know what they need to succeed. The same is true even for program directors and faculty facilitators, because their trainee population is heterogeneous, making it difficult to know what content would be most valuable. Moreover, it is not possible for any single individual to have the experience and knowledge necessary to teach all aspects of leadership — making it important to identify others who can do so. In this sense, some individuals have faith in the knowledge of others and are willing to ask for broad but ill-defined assistance, leaving it up to the discretion of URBEST to guide them with what they might want to deliver to their trainees. In fact, when first presented with the Marshmallow Challenge activity, the seminar director didn't think the activity was going to work, but we assured them that the URBEST trainees had selected this as their favorite activity. It certainly helped that the director witnessed the activity firsthand and heard students speaking positively about the session afterward in the hallways. That courage to trust is not, however, a given — it must be earned. As a result, it is important to advertise workshops, lessons, and even distributive learning units to program directors and coordinators within the university and centrally on the national BEST website.

Discussion and conclusion

Biomedical and STEM research training programs are currently in transformation, conscientiously evolving to prepare next-generation researchers for the full diversity of careers that they will ultimately experience in the scientific workforce, including in academia. Through URBEST, we have been able to introduce leadership and management domains as part of new curricula to our graduate students and postdoctoral fellows. We have included three different forms of learning — intensive, distributive, and on demand — to meet the needs of individual learners who have very different learning requirements and time availabilities. Each type of programming works to accomplish the same mission: develop future managers and leaders by helping them to discover and build leadership qualities and interpersonal skills that work for them. Introducing these practices to graduate students and postdoctoral associates is particularly beneficial to prepare future faculty members, as well as those future leaders who will select to pursue careers outside of academia.

Our first two forms of training — intensive and distributive — were very much comprehensive strategies with a trainee-oriented focus in their derivation and design, intended to develop trainees into proficient managers and leaders. Through our course evaluations, we observed that our intensive learning helped graduate students and postdoctoral fellows to discover and build leadership qualities and soft skills, while distributive learning made them aware of their intuitive skills, deficiencies they might have, and leaders within their institution who might be able to guide them with their personal growth. Intensive programming, such as is provided in *Leadership and Management for Scientists*, can be considered foundational and fits into the traditional method of instruction found at most institutions: the semester-long course. Although we are exploring alternate pedagogy for delivery, foundational courses are still a vital component of our preparation for our learners. It allows for depth in exploration of content and the ability to leverage a learning community that forms over time. Distributive learning, such as is presented in *Avoiding the Kiss of Death*, can be considered as an adaptive curricular component that can be integrated with other educational infrastructures. This allows the curriculum to be responsive and piloted with multiple learner groups. Both of these delivery styles should be implemented with broad input from students as well as institutional leaders. In our case, students are often the most important voice in the design and refinement of course content. They become more confident in guiding distributive programming after participating in one cycle because they see and understand the concept and are primed to be on the lookout for instructors to join the programming.

On-demand programming is very different from these other modes of training in that it is generated in response to a specific request, generally from a faculty colleague who seeks to meet the perceived needs of a trainee population for whom they have responsibility. Thus, the specific ask is completely dependent on the needs of that individual external to URBEST or of the population that the individual serves. Sometimes the ask is simply, *"We need some type of training that will help our students be more likely to succeed in their graduate work and future career."* This programming can be considered as "seizing the opportunity," as it is beneficial for the external individual but also results in potential programming that then becomes available for future use or adaptation by URBEST.

In essence, the programming, which we are demonstrating can be embedded in multiple ways, pushes the graduate student and postdoctoral culture toward leadership. Most notably, because changes occur based on asks from learners and faculty, we can use this accepting environment to help guide and establish a transformational culture of scientist leaders. In essence, we are at a *tipping point* [16]. Now is the time to meet the emerging needs for leadership and management education and guide the transformation of graduate education to include multiple forms and styles of learning. To develop into leaders, our students and postdocs need more than coursework. They need to be in a culture that leverages strong leadership and that transforms itself to meet these emerging needs.

We hypothesize that three things need to happen for the culture transition to be attainable and successful. First, universities need to "unlock" their non-academic units to serve as venues for the education of graduate students and postdocs. By using experts who play a leadership role within institutional resources (e.g., technology transfer and clinical research offices, regulatory oversight committees, etc.), learners will become familiar with the non-academic infrastructure needed for an academically strong university, and will meet role-model leaders they might have never met if all their learning had happened in a laboratory. In addition, unlocking non-academic units also facilitates interactions between learners and senior leadership who are often hidden in the open. Learners may already interact with some of these senior leaders in their capacity as scientists in their research laboratories, but they typically do not hear about the roles these senior leaders play in supporting and advancing the university mission. By using our social capital as educational administrators and leaders we can reveal these role models to biomedical and STEM graduate students and postdocs early in their career development, such that trainees have an opportunity to try on different leadership styles and build leadership qualities and interpersonal skills that work for them. An added value of these "unlocks" and "reveals" is in building connections across academic and non-academic units within the university, strengthening culture transition. With these connections, graduate students and fellows broaden the scope of their mentor network to include mentors who may develop into future advisors and influencers throughout their career.

Second, in helping graduate students and postdocs to discover and build leadership qualities and interpersonal skills that work for them, programming needs to move away from lecture-style sessions toward experiential and active learning strategies. This can be accomplished by incorporating interactive activities, reflective writing, self-assessment instruments, group discussions, and written instructor feedback. We found that our experiential and active learning strategies benefited trainees the most when learning took place in a structured (yet non-judgmental) environment where instructors believed that leadership styles were highly individual and situational and who did not perceive any particular styles as "good" or "bad." This style of instruction was sometimes novel to new and even experienced instructors; space and trust was essential so that instructors could get a feel for their students and fellow audiences and so they could reflect on and use their classroom experiences to further develop their content. In essence, the course helped some instructors explore a style of teaching with which they were not very familiar.

For example, one topic of *Leadership and Management for Scientists* that received limited feedback and praise in the first year of student evaluations is now one of our most popular topics and has been expanded from one to two sessions. For the first year, learners had pre-class readings, a short lecture, and group discussions about the content during a single class session. Outside of class, the individual learners worked on three short worksheets that needed to be completed as if they were PIs of their own lab. Because the topic was so unfamiliar to the learners, one class session did not provide enough time for the group to synthesize the information presented to them and move from half-formed thoughts to cohesive ideas. Because the students realized the PI role-playing assignment was valuable for their next real-life role, they were frustrated they could not engage in the task at a deeper level. Thus, they didn't want to eliminate the session; they wanted to deepen it. This was accomplished by expanding the topic into two sessions and by using one of the short worksheets as an in-class group exercise. The two sessions provided the students more time to grapple with the information and with a security in knowing they would meet with the instructor the following week for further discussion and refinement of their understanding, which they could then use to complete the final two worksheets.

Finally, and probably the most difficult requirement, is to build trust within the learner and university community around both the content domain of leadership development as well as the novel and possibly risky delivery methods. We have found that trust is not easily given, but must be earned. For us, trust was slowly built by providing programming we thought would appeal to the needs of our learners, accepting opportunities to develop programming to fit direct asks and needs of our learners and faculty, listening to the positive and negative evaluations (and hallway stories) shared with us, and refining accordingly. We also built trust by incorporating experiential learning techniques and through the direct interactions of trainees with multiple instructors, all of whom supported the idea of establishing a leadership culture in graduate student and postdoctoral fellow training. After three years, we feel we are starting to observe a shift both in the expectations of our students, trainees, and faculty in regards to leadership development, as well as within the scientific community and culture in wanting to support the growth and development of the next generation of leaders.

Looking to the future, we want to build on the success of our current delivery formats — intensive, distributive, and on demand — and begin to think of a developmental model along the training continuum of our trainee scientists. Learning how to be a leader in science is a complex task and requires layers of learning over time. A first-year graduate student might be best served to think about their own strengths and weaknesses as a leader, whereas a postdoc might benefit more from learning about conflict management and strategic planning initiatives. By developing a competency-based, developmental model for leadership, we can strengthen each layer in our scientific community and prepare our learners to understand and transition into leadership roles in the biomedical research workforce in academia or other sectors. We hope that — by introducing curricula, teaching in novel ways, and shifting culture — we can embed leadership development from the first days of *Ph.D.* graduate training and foster continuous development so that a leadership mindset naturally spans entire careers.

Acknowledgments

URBEST was funded by the Director's Funds of NIH to the University of Rochester, grant number 5DP7OD020315.

References

[1] Hill L. Becoming a manager: how new managers master the challenges of leadership. 2nd ed. Boston, MA: Harvard Business Press; 2003.

[2] Ciampa EJ, et al. A workshop on leadership for MD/PhD students. Medical Education Online 2011;16:7075. https://doi.org/10.3402/meo.v16i0.707.

[3] Hund AK, et al. Transforming mentorship in STEM by training scientists to be better leaders. Academic Practice in Ecology and Evolution 2018;8:9962−74.

[4] Van Noorden R. Some hard numbers on science's leadership problems. Nature 2018;557:294−6.

[5] Antes AL, et al. Are leadership and management skills essential for good research? an interview study of genetic researchers. Journal of Empirical Research on Human Research Ethics 2016;11:408−23.

[6] Bennett LM, Gadlin H, Levine-Finley S. Collaboration and team science: a field guide. Bethesda, MD: National Institutes of Health; 2010.

[7] Cohen CM, Cohen SL. Lab dynamics: management and leadership skills for scientists. 2nd ed. Cold Spring Harbor, NY: Cold Spring Harbor Laboratory Press; 2012.

[8] Gray B. Enhancing transdisciplinary research through collaborative leadership. American Journal of Preventative Medicine 2008;35(2, Suppl. 1):S124−32.

[9] Reis A, et al. Retention of junior faculty in academic medicine at the University of California, San Diego. Academic Medicine 2009;84:37−41.

[10] Ullrich L, et al. From student to steward: the interdisciplinary program in neuroscience at Georgetown University as a case study in professional development during doctoral training. Medical Education Online 2014;19:22623. https://doi.org/10.3402/meo.v19.22623.

[11] Mumford MD. Leading creative people: orchestrating expertise and relationships. The Leadership Quarterly 2002;13:705−50.

[12] Dunlosky J, et al. Improving students' learning with effective learning techniques: promising directions from cognitive and educational psychology. Psychological Science in the Public Interest 2013;(14):4−58.

[13] Kang SHK. Spaced repetition promotes efficient and effective learning: policy implications for instruction. Policy Insights from the Behavioral and Brain Sciences 2016;3:12−9.

[14] Individuals in the URBEST program participate in a variety of year-round offerings related to professional development and career exploration, which often includes discussions about leadership. Due to space constraints, we are only discussing our intensive, distributive, and on-demand offerings. For more information on our other offerings, visit our website at: https://urbest.urmc.edu.

[15] Wujec T. Build a tower, build a team. TED 2010. 2010. Available from: www.ted.com/talks/tom_wujec_build_a_tower.

[16] Fuhrmann CN. Enhancing graduate and postdoctoral education to create a sustainable biomedical workforce. Human Gene Therapy 2016;27:871−9.

Rutgers University's interdisciplinary Job Opportunities for Biomedical Scientists (iJOBS) Program: iNQUIRE, iNITIATE, iMPLEMENT, iNSTRUCT

Susan R. Engelhardt[a], Janet Alder[b]

[a]Center for Innovative Ventures of Emerging Technologies, Department of Biomedical Engineering, Rutgers University, Piscataway, NJ, United States; [b]School of Graduate Studies, Rutgers University, Piscataway, NJ, United States.

The iJOBS Program begins: We join the NIH BEST consortium

In 2014, Rutgers University became one of 17 schools in the country to be awarded an NIH Broadening Experiences in Scientific Training (BEST) grant. In September 2014, Rutgers's interdisciplinary Job Opportunities for Biomedical Scientists (iJOBS) Program was established to educate, mentor, and guide the university's 1,450 Ph.D. students and 620 postdoctoral fellows in the biomedical and life sciences to pursue skill-appropriate professional careers. Our mission was to empower our

 iJOBS Program blog

My journey begins

"You should definitely pursue your Ph.D.," they said. "You have the intellectual curiosity to succeed, the tenacity to persevere, and the passion to innovate!" … and so, I did. Upon earning my master's degree in biomedical engineering, I made one of the most important decisions of my life: I leapt into a Ph.D. program at Rutgers University, an institution dedicated to cutting-edge research, interdisciplinary collaboration, and professional development. In

BEST: Implementing Career Development Activities for Biomedical Research Trainees
https://doi.org/10.1016/B978-0-12-820759-8.00007-3

trainees to take action toward career development and build the knowledge, professional networks, and experience needed to transition efficiently to a wide range of careers.

As the iJOBS Program matured, we honed our focus on specific challenges facing our pre-doctorate students and postdoctoral fellows, collectively called "iJOBS trainees." Here, we present a program summary with key findings. However, to discuss our programming without the backdrop of a trainee's journey would leave out important context, so we have included a blog written from the perspective of a fictional trainee that describes their experience with our programming. All interpretations are based upon trainee feedback gleaned from annual program surveys, blog entries, and event-specific surveys.

iJOBS Program goals

Our initial goals were two-fold. We aspired not only to prepare our trainees to secure careers in the biosciences, but also to formally assess supply and demand for these fellows and to customize our programming to industrial/employer needs. In our annual iJOBS trainee surveys we learned that, on average, between 40% and 55% of our trainees felt that the iJOBS Program had, in fact, piqued their interest to pursue careers in industry, underscoring the fact that professional skills awareness and associated preparedness are our chief priority. Throughout the years, we have remained largely focused on day-to-day, holistic trainee programming. As a result of this focus, 53% of trainees surveyed in 2018 felt strongly that their knowledge of non-academic career options had increased.

iJOBS Program blog   *(cont'd)*

good company with over 5,200 student peers enrolled in more than 150 graduate programs across 3 campuses — New Brunswick/Piscataway, Newark, and Camden, NJ — I was excited to develop into a scientist who could contribute new knowledge to the biomedical disciplines through creative research and scholarship. I'd have the opportunity to learn from peers in so many disciplines, and through coursework and research training, I'd be prepared to advance the frontiers of the biomedical sciences.

It all seemed so noble; my head was "all in," but somewhere, in all the inertia, I had questions. Earning a Ph.D. degree is an enigma. It is a hard-earned rite of passage that brings one to the status of "scholar" following an arduous journey of disciplined research, authorship, and resulting growth. I was going to pursue my doctoral degree to extend my knowledge, to feed my love of science, and to focus on cell engineering, but I couldn't help but wonder what was waiting for me at the end of the journey.

When I embarked on this path, the obvious next step was my continued pursuit of research and, if I really stretched, I could see myself as a bench scientist within industry. As I progressed through my first year of doctoral studies, I confirmed my love of the biomedical sciences, their relevance to healthcare, and their potential societal impact. But a strange thing happened along the way. My focus shifted to the practical application of my research, and I faced my first Ph.D.-related existential crisis. I wondered, *If my research doesn't yield a positive impact on society, what is the intended purpose of all of my work?*

TABLE 7.1 Representative Rutgers University academic departments that participate in iJOBS.

Biochemistry, Biomedical Engineering, Cell & Developmental Biology, Cell Biology/Neuroscience/Physiology, Cellular & Molecular Pharmacology, Chemistry & Biochemical Engineering, Chemistry & Chemical Biology, Computational Biology & Molecular Biophysics, Endocrinology & Animal Biosciences, Environmental Science, Infection/Immunity/Inflammation Medicinal Chemistry, Microbiology & Molecular Genetics, Microbial Biology, Molecular Biology/Genetics/Cancer, Neuroscience, Nutritional Sciences, Pharmaceutical Science, Physiology & Integrative Biology, and Toxicology.

iJOBS Program reach

As initially proposed to the NIH, iJOBS was established as a partnership between Rutgers's Center for Innovative Ventures of Emerging Technologies (CIVET) and the School of Graduate Studies (SGS), and was established as a phase-based initiative. iJOBS was designed for current Ph.D. students, postdoctoral fellows, and recent alumni focused in biomedical/life sciences and engineering, addressing the needs of over 2,000 potential trainees spread across 27 distinct graduate programs and 32 departments (Table 7.1) in 3 geographically distinct, New Jersey-based, campuses (New Brunswick, Newark, and Camden).

As there was a diversity of departments under our iJOBS Program umbrella, the reinforcement of intra-campus academic relationships around common iJOBS goals and communication of our program's vision across campus boundaries were critical to our success. In addition to Rutgers-centric participation, a common NIH programmatic theme was inter-university collaboration; we invited students/fellows from our regional peer universities to participate in our programming. Students from iJOBS-affiliated universities — Princeton University, New Jersey Institute of Technology (NJIT), Stevens Institute of Technology, and Rowan University — benefitted from our programs, collectively making up 10% of our participating trainees.

iJOBS Program structure

Our program is organized into four phases (Fig. 7.1). Phase 1 programming, open to all pre- and postdoctoral trainees (as well as iJOBS Program alumni) from Rutgers and our affiliated universities, informs participants about potential careers and their requisite skills.

Phase 2 is application based and open to trainees who have attended at least 12 hours of Phase 1 programming, and provides trainees with opportunities for personalized learning experiences. Applications from trainees are received in May at all participating institutions and acceptances are communicated in June for the following program year. Trainees who are not accepted can re-apply to Phase 2 the following year. Trainee applications must include the following:

- Indication that the trainee has successfully completed the Ph.D. qualifying exam, or has a scheduled exam date.
- An unofficial transcript confirming that the trainee is in good academic standing (GPA > 3.0).
- An updated CV/resume.
- A personal statement addressing the trainee's iJOBS experience thus far, career goals (or options under consideration), skills, abilities, and perspective they hope to gain with continued participation in the iJOBS Program.
- Qualities and experience that the trainee brings to the iJOBS Program.
- Signed letter from the trainee's research laboratory principal investigator (PI) indicating their permission for the trainee's participation in Phase 2 of the program.
- Letter of recommendation from the trainee's PI (or another faculty member) describing the trainee's work ethic, independence, zest for learning,

 **iJOBS Program blog**

My journey clarified

When I began my journey toward earning a Ph.D., I couldn't see beyond the research and had few thoughts about my post-Ph.D. plans, (nor what I could even be with a Ph.D.). The iJOBS Program provided me with the perspective I needed and I attended many events during my first year.

My introduction to iJOBS programming was truly transformative. With support from my Ph.D. advisor and laboratory manager, I attended various panels, workshops, and mini-courses, and finally started to find my "professional voice."

Alongside a gathering of alumni representing a plethora of professions, I was privy to an insider's look at various career paths available to me at the one-day iJOBS Phase 1 workshop entitled "Non-Research Careers with a Ph.D." We explored three distinct career paths: teaching and education, government jobs, and non-research industry jobs and, despite the diversity of the profiled careers, there were common threads throughout the featured talks. The primary piece of advice was to get ourselves "out there" and apply for the jobs that interested us. Together, the panelists wove a message of hope, and almost demanded that we begin to value ourselves as the experts we were.

My next iJOBS stop was a workshop entitled "How to Pick the Right Career for You." It proved to be an enjoyable reflection on our individual "powers," each exercise flowing logically from one to the next as we identified our strengths, joys, and values to help us hone in on a suitable career choice. The workshop leaders stressed that we should begin our journies by considering career paths for which we were naturally suited.

I also attended programming that discussed bench research in pharma/biotech, clinical academic/professional acuity, and personal qualities, as well as the recommender's perspective regarding the trainee's future plans and potential.

Trainees that participated in Phase 2 are automatically accepted into iJOBS Phases 3 and 4. Phase 3 guides trainees through the job search process and Phase 4 is reflective, encouraging iJOBS alumni to serve as program mentors, host iJOBS programs, and contribute to the program at large. Each phase guides the trainee through the next decision point in their career search journey and serves as a demarcation of recommended introspection and activities: *What are my career options? What are the particulars of potential careers? How can I ready myself for such a career? How should I approach job placement?*

Our phased structure has proven itself to be a positive program feature. Although Phase 2 takes a year to com-

iJOBS Program blog *(cont'd)*

science, and regulatory science, among others. I decided that, given my penchant for data analysis, protocol, and order, and given my deep-seated desire to positively impact society with my life's work, I'd pursue the path of clinical operations. It seemed a perfect match. Although there are certain aspects of academia that I appreciate — intellectual freedom, sharing of knowledge with the next generation of scientists and engineers, collaboration and teamwork with colleagues and mentors, and work schedule flexibility — pursuing a career in academia was unlikely to fulfill my desire to directly impact patient health outcomes. So off I went to learn more about clinical operations and to learn the requisite tools to be accepted into, and compete successfully in, such a professional environment.

plete, iJOBS trainee participation in Phases 1, 3 and 4 can span multiple years. As trainees transition from one phase to the next, they meet phase-specific expectations and become increasingly accountable for their professional journeys. Trainees advance through the program phases based upon their individual growth and career readiness, although they typically delay their progression to Phase 3 until about a year before their pre- or post-doctoral work is scheduled to end so that it coincides with their focus on job searching. Once a trainee completes Phases 1—3, they may participate/contribute as alumni at any time.

iJOBS Program team

iJOBS Program leadership consists of a team of investigators and directors from graduate-level biomedical engineering and life sciences disciplines across all campuses, all of whom share responsibility for overall program administration and governance and who work with the extended team toward program design, implementation, and assessment. In turn, the extended team is made up of faculty program ambassadors and industrial partners that span all areas of the biomedical and life sciences.

Our program's diverse group of industrial partners facilitate opportunities for trainee site visits, shadowing (externships), and professional mentorship and inform iJOBS of skill requirements for professionals entering a career in their respective fields. Importantly, as New Jersey (NJ) is one of the top three biopharmaceutical clusters in the country, we have built ties to life sciences-focused trade organizations who have partnered with us to enhance the climate for life

Phase 1: iNQUIRE

THIS STARTING POINT FOR ALL TRAINEES
LAYS THE FOUNDATION FOR PROFESSIONAL
PREPAREDNESS AND PROVIDES INSIGHT INTO
AVAILABLE CAREER CHOICES.

Trainee entry experiences include career panels and workshops that inform business, management and communication skills essential to career success. Industrial partners continue trainee dialogue with hosting of site visits and participation in networking sessions.

Phase 3: iMPLEMENT

TRAINEES PREPARE FOR SUCCESS WITH
INDIVIDUALIZED GUIDANCE DURING PROFESSIONAL
SEARCH AND PLACEMENT.

Trainees communicate their skills and expertise with resume and application package development and hone their interview skills with Rutgers iJOBS training and coaching. Focused career fairs showcase suitable career opportunities and provide industrial partners access to these professionally competent candidates, many of whom they have guided and mentored.

Phase 2: iNITIATE

TRAINEES COMMIT TO THE iJOBS PROGRAM AND APPLY
TO TAKE PART IN INTENSIVE TRAINING FOR THEIR
CAREER TRACK OF CHOICE.

Having selected a specific career track, each trainee attends an academic course in his/her focus area to build professional perspective and skill. To supplement the academic experience, Rutgers iJOBS industrial partners demonstrate the practical side of these careers while serving as trainee mentors and shadow partners. Throughout, individual development plans document the journey to encourage trainee advancement toward their career goals.

Phase 4: iNSTRUCT

PROGRAM ALUMNI ARE ENCOURAGED TO SHARE
THEIR WISDOM WITH NEW iJOBS TRAINEES.

Training is insufficient without the benefit of experience. Upon embarking upon a professional career, program alumni are encouraged to share their wisdom with new trainees and serve as mentors, event hosts and shadow partners.

FIG. 7.1   Rutgers University's iJOBS Program phases.

sciences in NJ. These partnerships have also helped us to advertise the iJOBS Program to potential industry partners, host iJOBS events, and garner program feedback. One example is our joint programming with BioNJ (a NJ-based trade organization focused on enhancing the climate for life sciences in the state), where we partner to connect NJ-based companies seeking to hire doctoral students, while providing tools to facilitate better access to iJOBS's Ph.D. talent pipeline.

Areas of career focus

Back in 2014, with funding in hand and our partners oriented and ready, we began to build our program. We reviewed our proposed career areas of focus and considered programming options appropriate to each one. We began our journey with five focus areas (Fig. 7.2), but as we implemented the program, we found that real-life careers were not limited to these five areas and that there were nuances in each that gave rise to additional career thrusts. Our new career areas include (but are not limited to): (A) bench research in government, pharma, or biotechnology; (B) teaching-intensive careers in academia, specifically in Primarily Undergraduate Institutions (PUIs); (C) clinical research; (D) regulatory affairs; (E) business consulting; (F) scientific writing and medical communications; (G) medical affairs; (H) work with non-profit organizations and foundations; (I) finance and equity research; (J) publishing; (K) food safety; (L) journalism; (M) teaching education outreach; and (N) entrepreneurship.

Trainee resourcefulness expanded as we progressed, and we expanded our program alongside them. **We learned that our iJOBS trainees, armed with scientific, engineering, and professional skill and prowess, could take their perspective to any professional setting and flourish.** We learned to keep our "career destinations" flexible and are proud to note that our program has helped our trainees imagine and pursue limitless opportunities.

Program communication

 **i J O B S P r o g r a m b l o g**

My growing skills toolbox

I finally started the third year of my Ph.D. studies. Although I continued on my professional development journey, the pursuit of my Ph.D. took precedence and my days were replete with conducting research, reviewing scientific literature, and honing my lab skills (techniques, equipment, data analysis, etc.).

I realized that, to be competitive for the professional positions I would seek upon earning my degree, I required additional training with a focus that went beyond what I could learn in my research lab. Although work in the lab continued to refine my communications skills, I had to learn how to communicate appropriately across functional roles typically found in industry. Academic courses that touched upon critical thinking and experimentation design were adequate for academic research, but I also needed didactic exposure to business principles, with practical, team-based casework aimed at solving real business problems. Although my research team's work in the lab and my team- and case-based courses taught me to collaborate effectively and to successfully manage research projects, learning to do so within a professional environment was critical to my growth and professional preparedness. Finally, my technical knowledge and the principles that guided me were limited to my discipline of study, peppered with subjective learnings that benefitted my research, but I needed more to succeed.

iJOBS offered a SciPhD certificate program called, "The Business of Science: Applying the Scientific Method to Succeed in Industry," so I

We reached out to our first trainee pool in various ways: presenting at departmental faculty meetings and departmental gatherings, inserting our program details into academic orientations, launching a website [1], garnering publicity through news articles and press releases, starting a student-led blog [2], and sending email upon email that we were "open for business."

We update our programming and advertise our scheduled events weekly with the help of an online event scheduling system. iJOBS event registration is customized to include mini-surveys so that we can analyze the demographics of our attendees. We also take advantage of social media, and blog, email, and tweet to maximize the number of prospective participants. We always create downloadable summaries (in the form of podcasts) of the latest events and post them on our website, preserving the information for posterity and allowing it to serve as reference for interested trainees and/or alumni. **Given programmatic reporting requirements and required communication to a vast trainee and partner audience, a proactive communications strategy has been critical to the success of the iJOBS Program.**

Sixty-two percent of our trainees, surveyed on an annual basis, noted that email communication is the most effective method of communicating with them, followed not so closely by faculty encouragement at 16% and word of mouth at 10%. Therefore, we continue to

iJOBS Program blog *(cont'd)*

signed up and off I went in search of more perspective. I joined nearly 60 graduate students and postdocs for this 4-day workshop to learn how to frame skills gained in academia to be marketable toward careers in industry. We assessed our personalities and put ourselves in each other's shoes to be better able to navigate conflicting personas and team relationships. We explored team building and business networking using physical group activities that encouraged us to work as a unit and strive toward innovation, creativity, and better communication. We learned firsthand — while standing in a circle and throwing around random objects! — that effective communication and project management are key factors in nurturing innovation and success. The assignments challenged us to build biotechnology companies, create process flowcharts, draft business plans, and pitch it all to our iJOBS peers. An introduction to the world of finance followed, and we were assigned roles and tasked with inter-organizational deal-making to manage our "businesses." We held mock job interviews daily and frequently received valuable feedback.

Whew! That whirlwind of information, perspective, and team-building was very much worthwhile, but more was yet to come!

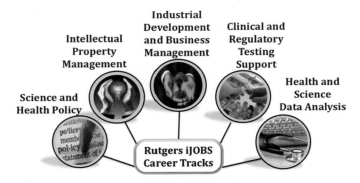

FIG. 7.2   Initial Rutgers iJOBS career tracks, circa 2015.

utilize email as our primary vehicle for iJOBS Program communication. **Best program practices include establishing and continuously updating email distribution lists that include program influencers and stakeholders, keeping our messaging tightly categorized and to the point, and setting a schedule for program emails so that recipients pay attention to updates.**

iJOBS programming

With our first cohort in tow, our charge was to help the trainees identify not only what careers they *wanted* to pursue, but also which ones they *didn't* want to pursue and why. We researched the tenets of successful career journeys, referenced our professional networks, double-checked our alumni contact lists, gave a silent nod of thanks that NJ, known as "The Cure Corridor," is brimming with life sciences-focused industries, and began to develop our Phase 1 programming. Our program phasing followed the trainees' professional journey toward career placement (Table 7.2). Career panels evolved into job simulations, case studies,

TABLE 7.2   iJOBS's phased program elements.

Phase	Description	Elements
1	Trainees learn about professional tracks relevant to those with biomedical sciences doctoral degrees.	• Weekly iJOBS events featuring career panels, simulations, seminars and workshops highlighting professional options, required professional skills, and job search preparedness. • Site visits to partnering companies, where students are given operational tours and have an opportunity to interact with employees across various functional roles. • Quarterly networking events with professionals of interest. • Access to the iJOBS Program website portal, where trainees can garner professional insights, access related resources, etc. • Participation in the "SciPhD" course, taught by Human Workflows, LLC, which teaches trainees about communication, business skills, and frameworks required for professional success.
2	Trainees are educated in core professional skills and those relevant to their career paths.	• Assignment of a professional shadowing host selected based upon trainee career interest(s). • Attendance at an academic course in a trainee's career focus area to build professional perspective and skill. • Assignment of an external mentor with whom trainees can consult to help advance their specific career goals. • Trainee planning and tracking of professional progress in their Individual Development Plan (IDP).
3	Trainees prepare for professional placement.	• Resume and application package development. • LinkedIn profile development. • Interview skill enhancement/role-playing support. • Trainee and industrial partner participation in career fairs. • Assistance with job placement and tracking thereof.

(Continued)

TABLE 7.2   iJOBS's phased program elements.—cont'd

Phase	Description	Elements
4	iJOBS alumni volunteer to support iJOBS programming.	• Alumni access to all program resources. • Invitation to "give back to the program" by providing mentoring and shadowing support to current program trainees.

and interactive workshops as trainee involvement was essential for learning; workshops evolved into mini-courses when learning objectives were intensely skills based; and site visits became a favored event for all trainees. Programming examples are shown in Table 7.3.

TABLE 7.3   Rutgers University's iJOBS Phase 1: career exploration and job search skills programming.

Representative iJOBS Programming		
Career Exploration		
PANELS		**SEMINARS**
Bench research in pharma & biotech	Science to medicine forum	Public policy
Careers in biomedical STEM	Introduction to professional tracks	Science communications
Chemical biology research	Medical affairs	Science policy
Clinical & regulatory science	NIH government employment opportunities	Science publishing careers
Clinical diagnostics	Science storytelling	Scientific writing & journalism
Contract clinical research	FDA jobs/research positions	Non-faculty university jobs for Ph.D.s
Contract research organizations	Life after graduate school	Women in academic biology
Data science	Medical devices	**WORKSHOPS**
Education & science outreach	Non-profit organizations	What can you be with a Ph.D.
Finance & equity research	Non-research careers with a Ph.D.	Biopharma consulting case study
Grant writing	Pharma medical education	Pharma agency communications space
Health data analysis	Product development, technology & sales support	Biotech investment space case study
Intellectual Property Law		
Job Search Skills		
Career fair/networking preparation	Internship discovery, application & resume preparation	Resume/CV development
Choosing the right career	Interviewing skills & practice sessions	Seeking employment in the US (for international students)
Emotional intelligence for self-awareness & influence	Professional profiles: transferable skills & experience	Staffing agency job searching
Industry job application strategies	Professionalism tips for getting a job & keeping it	Strategic networking on- & offline to propel your career
Strengths & talent assessment		

Along the way, we built a few outstanding courses that were designed to further prepare our trainees for their professional journeys. Courses, which are grant (NIH or Burroughs Wellcome) sponsored, include "Communicating Science," which teaches students how to present to diverse audiences and clearly explain the significance and potential benefits of their research; "Professional Preparedness for Biotechnology," which introduces additional layers of specialized competence in areas of project, financial, operations, organizational, and risk management; and "From Molecules to Medicine," which provides students with a practical, hands-on look at drug development in the pharmaceutical and biotech industries.

We recognized the need for a dynamic mix of programmatic offerings relevant to each step in the trainees' career journeys, and realized that this mix needed to be replicated in all program locations, as our teachings were timeless and our audiences varied over time.

We exposed our trainees to leading-edge thinking relevant to all career foci, presented by our industrial partners, peer institutions, and trade associations, and provided our trainees with valuable perspectives on trends and current events related to career responsibilities. For example, topics explored in life sciences, biotechnology policy, and politics workshops included discussions on issues such as, "Opioid abuse: Searching for solutions in science and politics," "Advocating for science: How to inform and persuade politicians," "Implications of the zika virus on science policy," and "Communicating risk regarding science and health: Lead toxicity and public policy," amongst others.

In addition to understanding the "business and politics" of a specific career, we offered workshops to teach career-specific skills required by prospective employers and often listed in job requisitions. We addressed skills in areas such as business planning, computing skills for genomic research, good laboratory practice (GLP), pharmacokinetic/pharmacodynamic analyses, project management principles, regulatory writing, teaching techniques, and others. Many of these programs were facilitated by our program partners, to whom we are continually grateful.

Site visits were an important part of our programming. Sites visited include biologics, medical device, and pharmaceutical companies of all sizes; biotechnology incubators and enterprise development centers; government departmental offices; and advocacy companies. Owing to the generosity of our partners, we have been treated to facility tours and to interactions with professionals from all functional areas, providing trainees with an understanding of the corporate organizational structure and of functional contributions to the overall success of such companies or entities.

Below we offer a few tips for those who seek to set up their own site visits:

1. Ensure diversity of showcased facilities, organizational functions, and roles.
2. Ask the site visit host to discuss the company's culture with their trainee visitors.
3. Clarify to the visiting trainees that attending a site visit is not a guarantee of a follow-up job offer. Rather, communicate that site visits are group discussions with willing partners who will help the trainees garner professional perspective and context.
4. Be mindful of the partners' preferred mode of communication and don't share their contact information with the trainees without the partners' prior approval. LinkedIn is often the preferred mode of connection/communication (even over email).

5. Ensure that time for networking and one-on-one conversation between professionals and trainees is allotted during each site visit.

iJOBS programming provides trainees with vital professional skills, but they gain so much more. **Trainees become conversant in the jargon used in their professions of interest and thus can excel during networking sessions and job interviews. Importantly, they cross the threshold into employment with a basic vocabulary for, and a rounded perspective of, their chosen professions.**

Trainee shadowing, mentoring, and coursework

Acceptance into the application-based Phase 2 brings with it additional opportunities that allow each trainee to reflect on their career choices while gaining invaluable, real-world perspectives. Once accepted into Phase 2, trainees participate in a shadowing program wherein each trainee is matched with a partnering company, firm, or institution focused on work relevant to the trainee's career of interest. Shadowing, sometimes called an externship, is a close cousin to informational interviewing [3], but it takes place on a protracted scale. Shadowing through iJOBS provides trainees the opportunity to observe successful practitioners in their area(s) of interest and to learn exactly what pursuit of such a career path involves. This experience has a required commitment of 72 contact hours, satisfied in any way that is convenient to both the trainee and the shadow host, over the course of an academic semester. Nondisclosure agreements are negotiated and executed for all shadowing engagements to alleviate concerns related to trainee exposure to confidential information and inadvertent sharing of Rutgers's intellectual property.

In addition to shadowing, we provide each Phase 2 trainee with access to a professional mentor who is expected to fill in

 iJOBS Program blog

Immersion in the details and experience

During the third year of my graduate studies, I was accepted into iJOBS Phase 2, and I met with the program directors to discuss my journey and how I would utilize the additional programming offered.

Coursework. Shadowing. Mentorship. So many opportunities to observe, learn, and better understand the experiences of a clinical operations manager!

First stop: coursework. Phase 2 trainees are invited to register for a full, credit-bearing course in an area related to their career choice. It's an opportunity to become entrenched in the details while mingling with like-minded students and fully engaging with the subject matter. The choice of course was up to me, so I decided upon a Rutgers course entitled, "Fundamentals of Regulatory Affairs," during which I gained so much perspective regarding the laws, regulations, and regulatory agencies governing pharmaceuticals, devices, biologics, and combination products marketed in the US and in the world. I learned the historical context in which the FDA evolved, its structure and its relationship with other US regulatory agencies, and market clearance pathways for the products they work with to ensure the development and delivery of safe and effective healthcare products. I aced the any knowledge gaps with practical perspectives regarding careers of choice and the trainee's pursuit thereof. iJOBS mentors are selected to participate based upon their professional foci and experience, and they are expected to guide their assigned trainees and recommend requisite skills so that they are well prepared when they enter the workforce. Mentors impart wisdom, technical knowledge, assistance, and support to their mentee throughout — and often beyond — the iJOBS Program journey.

In order to get the most out of their shadowing and mentoring experience, trainees complete an (NIH required) Individual Development Plan (IDP), noting skills required for career success and logging their activities throughout. The logs include the following information:

- Date of activity.
- Influencing person and functional role: a record of a person that influenced the logged activity (could be shadow host, mentor, or someone within the hosts/trainee's network) and their functional/professional role.
- Observation and impact on career focus: What did the trainee observe and how did it influence their career focus? Example entries include developing an understanding of corporate and organizational policies, as observed during an inter-departmental dialogue, and of required work products, gleaned from shadow discussions.
- Implementation: What steps will the trainee take to integrate these observations/influences with their planned career journey?
- Documentation: How are these observations recorded in the trainee's Individual Development Plan?

iJOBS Program blog *(cont'd)*

class, owing to my passion for the subject and my quest to succeed.

Next stop: shadowing. Matched with a NJ-based pharmaceutical company focused on many therapeutic areas and advanced biotherapeutic and research platforms, I discussed the shadowing logistics with my host, cleared the schedule with my PI (with whom I'd discussed my absence from lab prior to reaching out to my host), and was set to begin. I wasn't sure what to expect: would I feel uncomfortable? Would I be expected to contribute in uncharted territory?

No worries. It proved to be an amazing experience. I visited two different company Research & Development sites and was fortunate to spend time with individuals involved in the drug development process, as well as with clinical statisticians, research scientists, and regulatory affairs scientists. Shadowing in oncology group business meetings focused on clinical studies taught me the value of outstanding communication skills and a passion for team-based work. My shadow hosts were generous with their time and gracious with their interactions, and were inexhaustible fonts of priceless encouragement.

Overall, my shadowing experience helped me to capitalize on this new perspective, improving my efficiency in the laboratory and increasing my productivity as a doctoral student.

As a Phase 2 iJOBS trainee, I was also matched with a mentor, a consultant driving the execution of global regulatory strategies at the company where I shadowed. My mentor helped me to identify requisite skills for career success, integrate them into my IDP, and work toward adding these skills to my "professional toolbox," serving as a perfect complement to the shadowing experience.

At the conclusion of each shadowing and mentoring engagement, the trainee, shadow host, and mentor are surveyed to measure the effectiveness of the experience.

Although the shadowing and mentoring experiences are dynamic and can't be predicted, iJOBS provides a primer with tips and guidelines for the trainee, shadow host, and mentor, to be referenced at their discretion. Representative trainee experiences are as follows:

- **Science and health policy:** Trainees observe public policy think tanks and/or public officials analyzing regulations to influence state-level legislation. They develop an appreciation of information analysis techniques, communication of research findings, consulting skills, and other such competencies.
- **Industrial development and business management of science:** Trainees observe professionals with varying functional roles as they participate in business meetings, team problem solving and strategic analyses. They develop an appreciation of compelling value propositions, competitive analyses, market segmentation, financial forecasting, business modeling, and go-to-market strategy development.
- **Intellectual property (IP) management:** Trainees observe professionals as they perform their functions relevant to IP management, marketing, and contract negotiation. They are exposed to the basics and the relevant laws and regulations behind IP management, develop an appreciation of the process of basic patentability searches and technology assessments, and begin to understand what is entailed in the commercialization of science-based innovation.
- **Clinical and regulatory sciences:** Trainees observe professionals involved in clinical trial management and glean perspective from both the sponsors' and the study sites' perspectives, gaining experience in protocol development, internal review board submissions, managing study budgets, facilitating proper data collection, preparing for study audits, and more. Trainees also observe strategies for patient recruitment and develop an understanding of the policies governing product development with regard to drug safety plans and risk management strategies.
- **Health and science data analysis:** Trainees observe professionals planning, developing, and maintaining health record storage and retrieval systems. They come to appreciate the process of importing, cleaning, validating or modeling data, and drawing relevant conclusions as these professionals prepare to present data for the design and development of relational databases and evaluate information to determine compliance with standards.

We were fortunate to engage with partners who were excited to serve as shadow hosts and mentors and to give back to the community by sharing their perspectives with our trainees. Year after year, these experiences have been productive for all parties involved, and our partners have continued to support our program and have asked to be matched with trainees in subsequent years. **Although it is admittedly convenient to rely on strong partners when setting up each new cohort for their own shadowing and mentoring experiences, it is also imperative to continually expand the partner network so that trainees have access to changing professional environments and perspectives. Logs and surveys are critical for the directorship to measure the trainee experience, to gain formal feedback, and to ensure that there is linkage between the trainees' experiences and actionable next steps documented in their IDPs.**

A course of choice

Each iJOBS Phase 2 trainee has the opportunity to choose and attend a three credit, iJOBS-sponsored, academic course at Rutgers in his/her focus area to build professional perspective and skill; some trainees, however, have opted to attend certificate programs or external workshops in lieu of formal classwork. Fig. 7.3 illustrates representative courses selected by trainees and participating Rutgers schools. The list of courses available to trainees, however, undergoes frequent changes as our trainees uncover new academic venues to learn the detail, culture, and cadence of their intended careers. Trainees find that taking an academic course is a challenging, but nevertheless rewarding, experience as they are exposed to topics, practices, and skill building otherwise not accessible to them within their Ph.D. journeys. **While enrolled in the course, trainees are advised and encouraged to mingle and extend their reach to students from other disciplines so that they can maximize their learning experience**.

From an operational perspective, program leadership must develop relationships with administrators of Rutgers's academic schools, departments, and programs to ensure a smooth process for trainee course registration, tuition monies transfer, and update of trainees' academic transcripts with the appropriate credits. iJOBS leadership secures the participation of schools or departments that offer these courses and confirms that they are willing to

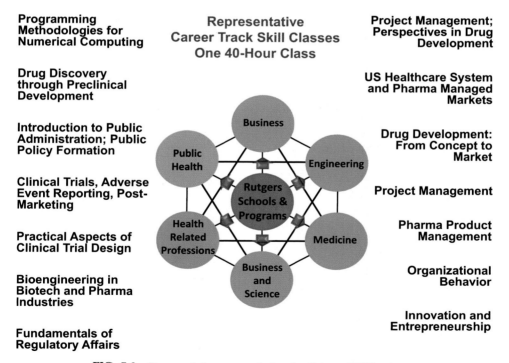

FIG. 7.3   Representative course choices for Rutgers iJOBS trainees.

have our trainees join their classes; for example, the Rutgers Business School must agree to enroll our trainees from the Rutgers School of Environmental and Biological Sciences into their courses. Occasionally, a trainee will require special handling if, for example, they choose to withdraw from a course if it doesn't meet their learning goals. **Strong collegial relationships help to weather inter-organizational concerns and keep the program on track.**

Communicate, communicate, communicate

The key to program success is timely and informed communication with stakeholders, built around formal communication touchstones. The NIH recognized this and required that program leaders participate in an annual BEST meeting, attended by all of the awardee team leaders. In addition to these yearly meetings, Rutgers holds biannual iJOBS Program symposia, scheduled in April and November of each program year, that provide us with opportunities to reconnect with our partners, recognize the accomplishments of our trainees, and work in groups to discuss how we can improve iJOBS programming. These symposia are scheduled six months in advance; all program partners (anyone who has hosted and/or mentored a trainee, held a site visit, or participated in a panel, workshop, or seminar) are invited to attend, as well our NJ-based, biosciences-focused trade organization partners that help keep our program current, relevant, and well- advertised. In addition, we invite our local affiliate universities (Princeton University, New Jersey Institute of Technology, Stevens Institute of Technology, and Rowan University), as well as universities outside of NJ that express interest in the BEST program, to share their academic perspectives. **All Phase 2 iJOBS trainees are**

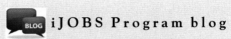 **iJOBS Program blog**

Ready, Set, Go Find a Job!

My third year as an iJOBS trainee (and my fourth as a graduate student) had begun and I was scheduled to defend my thesis in 10 months. The clock was ticking. Faced with navigating the bridge between career exploration and job placement, I prepared myself for the challenge and looked forward to Phase 3 iJOBS programming to guide my journey.

Thanks to the iJOBS workshops that I attended in Phase 1, my resume and CV were ready for prime time, but I was not confident about my ability to recognize a fitting job requisition, apply the appropriate strategy in filling out the application, and ensure that my cover letter underscored my fit for such a position.

Thankfully, there was a plethora of resources that I could make use of! Our program's partners offered amazing services:

- 2Actify: LinkedIn profile development modules, tips for strategic online networking, and one-on-one consultation sessions.
- Hart & Chin Associates, Oystir, and DOC: Two private sessions for resume and cover letter review, as well as interview preparation.
- BioNJ: Job listings for iJOBS trainees of NJ-based companies looking specifically for newly-minted Ph.D.'s and listings of job-seeking iJOBS trainee profiles for these companies to access our biomedical sciences Ph.D. talent pipeline.

expected to participate in these symposia as they serve as program mileposts and are important for program cadence, communication of latest updates, and in-person networking.

Future iJOBS programming: "Survey says …"

How can we improve our programming in the future? To help answer this question, we survey the trainees annually to glean valuable recommendations, examples of which are summarized below (in no particular order):

- Align better with graduate schedules/availability for programming and include family-friendly scheduling.
- Continue to offer iJOBS events at all three Rutgers locations.
- Expand iJOBS's career focus to those with dual degrees (M.D./Ph.D., Pharm.D./Ph.D., etc.).
- Increase collaborations between academic labs and industry partners so that graduate students can work with industry as part of their formal training.
- Continue to offer additional interactive workshops with group learning activities.
- More closely engage Rutgers' subject matter experts in business, law, and other relevant disciplines for expanded programming.

iJOBS Program blog *(cont'd)*

To assist in my job search, I attended a workshop entitled, "How to Prepare for the Industry Job Market," where I met with corporate Human Resources representatives and discussed the best ways to apply for jobs at pharmaceutical companies of various sizes and foci. I also picked up a few tips and tricks such as the importance of having a strong social media presence, posting a professional photograph to my online profiles, and creating a robust LinkedIn profile using keywords that highlight my experience.

I held my breath and began to respond to job requisitions of interest. My efforts yielded three interviews with large pharmaceutical companies, two of which were located in NJ and one that would take me to Toronto, Canada. Although I was most intrigued by the Canadian opportunity, I decided to stay local and was offered a position as a Clinical Research Coordinator. Not my final destination, but a proper first stop that I was thrilled to have secured.

I will, of course, give back to iJOBS and serve as a shadow host and mentor to future trainees embarking on the path that I have just traversed. I might even return to campus, attend a workshop or two and take the opportunity to mingle again with the program that enabled me to define my professional interest, present myself in the most informed and capable way, and secure *an incredible professional opportunity!*

Signing off for now, but I'll be in touch again soon!

Of course, we benchmark with our peer BEST programs for continuous program improvement. **We gained a lot from the annual NIH-hosted BEST conferences, where pearls of wisdom were consistently shared across all 17 BEST member institutions and where cross-member working teams spontaneously formed to review and report upon important program-related matters.**

Words of advice for potential program leaders

- Follow your instincts and allow your trainees to lead the way. Resist stifling trainee creativity with a rigid program framework. Be flexible, but remain mindful of repeatable processes.
- Leverage existing resources at academic schools, institutes, and centers.
- Advertise trainee successes, as it encourages continued greatness.
- Keep your alumni involved and close to the program; their perspective is invaluable and their contributions priceless.
- Feed and nurture your partner relationships; they are the lifeblood of your program.
- Communicate, communicate, and communicate to all of your stakeholders, from executive supporters and industrial partners to trainees and everyone in between. These relationships are the foundation of a successful program.
- Finally, let your program's reputation inform its success and its growth.

Parting words

The Rutgers iJOBS Program has been deemed a veritable success across all of our campuses and across our participating organizations. It has become almost a sub-culture amongst our doctorate students, postdoctoral fellows, and alumni, and it is acknowledged as a "must-have experience" that many of our prospective Ph.D. students recognize as a major draw to our university. Our professional partners understand the value of iJOBS and remain committed to the program as our trainees cross the threshold from academia to professional careers with much more than just an inkling of what they aspire to be, ranking them above the competition. We have made a difference in these trainees' professional lives and delivered motivated, skilled, and informed doctorate fellows into the biosciences workplace, where each one of them is sure to have a positive impact.

Acknowledgments

Rutgers University's iJOBS team thanks the NIH BEST initiative for including us amongst the 17 universities given the opportunity to build this amazing program (iJOBS; 5DP70D020314). It was an invigorating experience to be a part of the extended "BEST" team, benefitting from the benchmarks of best-in-class programming and the camaraderie that made it all accessible! Special thanks to the Rutgers School of Graduate Studies, the Office of the Senior Vice President for Academic Affairs, and the Office of Research and Economic Development for championing and supporting our program throughout our five years of iJOBS program success. A resounding thank you to our program partners in industry, government, trade/economic associations, and non-profits who were the wind beneath our wings, demonstrating to our trainees what each one of them could be with a Ph.D. in the life and biomedical sciences. Finally, thank you to the entire iJOBS team for their efforts over the years, including principal investigators James Millonig, Ph.D., and Martin Yarmush, M.D., Ph.D.; program directors Janet Alder, Ph.D., Doreen Badheka, Ph.D., and Susan Engelhardt, M.S.; and program manager Sunita Chaudhary, Ph.D.

References

[1] Rutgers interdisciplinary Job Opportunities for Biomedical Scientists, n.d., Available from: http://ijobs.rutgers.edu.

[2] Rutgers interdisciplinary Job Opportunities for Biomedical Scientists blog, n.d. Available from: http://ijobs.rutgers.edu/blog.php.

[3] Informational interviewing is a career-exploration or job-seeking tool where interested parties conduct short ($\sim$30-minute) interviews with people in their prospective professions to learn more about those fields.

The FUTURE partner network: Linking graduate students and postdoctoral scholars to Ph.D. professionals to inform career decisions

Jennifer Greenier, Stacy Hayashi, Bineti Vitta, Rachel L. Reeves, Milagros Copara, Daniel Moglen, Frederick Meyers, Lars Berglund

Clinical and Translational Science Center, University of California, Davis, CA, United States

Introduction

FUTURE is a career exploration program at the University of California, Davis, that was established with funding from the NIH BEST Initiative. The program is administered by the Clinical and Translational Science Center (CTSC) and is now fully supported by the CTSC and the School of Medicine. FUTURE is open to predoctoral graduate students and postdoctoral scholars conducting health-related research across 20 graduate programs and 29 departments in two campuses. Thus far, 131 participants in seven cohorts have enrolled in the program, representing the Schools of Medicine and Veterinary Medicine, and the Colleges of Biological Sciences, Agricultural and Environmental Science, Letters and Science, and Engineering. The centerpiece of the FUTURE program is a 10-week Career Exploration workshop series, which serves as a foundational career preparation experience for newly enrolled participants and as their introduction to a unique feature of our program, the FUTURE Partner Network. The Partner Network comprises Ph.D. professionals in non-faculty careers who contribute to the FUTURE program by sharing their career knowledge and experience with

BEST: Implementing Career Development Activities for Biomedical Research Trainees
https://doi.org/10.1016/B978-0-12-820759-8.00008-5

program participants. In this chapter, we'll describe the network, our data management methods, and the interactions between our program partners and participants.

From the inception of the FUTURE program, we recognized that although many trainees know that they do not want to pursue faculty research careers, they have limited knowledge of the diversity of options that are well suited for Ph.D.-trained scientists. Even in cases where they do know about particular careers, trainees often lack knowledge about how to transition into them [1]. Successful transition of Ph.D. trainees into careers of their choice requires insider information about what various careers entail, potential paths to those careers, and the skills needed to pursue them. One way of addressing this need is to connect graduate students and postdocs to Ph.D. scientists working in a broad spectrum of job categories in all major career sectors who can provide insight into interesting and fulfilling careers inside and outside of academia.

Our vision of a career preparation program for graduate students and postdocs included a network of Ph.D. scientists who had successfully navigated non-faculty career paths and who could provide trainees with windows into a variety of professions, from the known and expected to the unknown and unimagined. We aspired to build a community of Ph.D.s in the workforce who would partner with us in this effort, and eventually grow to a critical mass of professionals who adequately represented the diverse career interests of our program participants. We began recruiting partners in the second year of our program, primarily from our own professional contacts and from career-focused campus events, all while experimenting with different strategies for developing and managing our network. By the end of year 4, nearly 100 Ph.D. scientists had joined our FUTURE Partner Network, and we had refined our processes for recruitment, data management, and partner engagement.

Partner recruitment

Our primary recruitment strategy is to use LinkedIn to reach out to UC Davis Ph.D. alumni working in career areas of potential interest to our FUTURE members. Referral of Ph.D. colleagues by current partners accounts for our second most common method of recruitment, followed by direct in-person contact with potential partners at career-focused events for graduate students and postdocs. A growing source of new recruits is our community of FUTURE program alumni who have transitioned to their own careers; several have now rejoined the program as partners in our network. We also recruit partners through our website [2], where we provide details for those who are interested in joining the network.

Network composition

Currently, our partner network consists of 106 Ph.D. professionals in careers representing various workforce sectors and common job functions [3] for biomedical scientists in non-faculty positions (Fig. 8.1). The majority (67%) of our partners are UC Davis alumni. Although 18% of our partners reside out of state, our network is mostly regional, with 37% concentrated within a 60-mile radius of UC Davis and 82% located within 100 miles.

 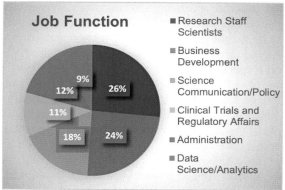

FIG. 8.1   FUTURE partner network members (n = 106) represented by workforce sector and job function.

As most (75%) of our network members are within 10 years of completing their doctoral degrees and many (36%) are within five years of earning their Ph.D.s, their experiences reflect current job market trends and career trajectories and are therefore highly relevant to our trainees who are on the cusp of entering the workforce.

Partner data management

A growing and evolving partner network requires an effective database management system. We first tried maintaining our data using a well-known customer relationship management platform, but it proved to be too complex and not flexible enough for our needs. We then moved to a spreadsheet format, but it did not facilitate team collaboration and quickly became unwieldy as our database grew. We finally settled on a cloud-based data management service that is a database/spreadsheet hybrid and that enables us to collaborate efficiently, sort and filter data, and link related fields across tables. We use our database to store information about our partners, including partner name, degree, company/organization, location, email address, career sector, and job function, and to track partner engagement with our program (e.g., through activities and interactions with our participants). The ease of data retrieval from this platform greatly facilitates the preparation of reports and the dissemination of information about our partner network.

Partner engagement

Although our minimum requirement for membership in the partner network is a commitment to conducting at least one informational interview per quarter (up to four per year), some partners may not be matched with trainees every quarter and some may volunteer beyond the minimum number of interviews. Most of our partners are interested in engaging with the program beyond informational interviews, however. Several options for additional engagement are offered: participation in program activities such as workshops, career panels, and networking events; hosting trainees for job shadows, site visits, or internships; and volunteering to be a career mentor to one or more program participants. The two most common means of engaging partners in program activities are through monthly small-group discussions focused on a specific partner's career path and position, and through our career exploration workshop series, where partners participate on career panels or lead discussions about various career-related topics, such as negotiation, social media tools, networking, and mentorship.

In 2018, 129 partner-participant connections were facilitated by FUTURE, resulting in 96 informational interviews and numerous interactions through career panel presentations and workshop appearances. In addition to volunteering for informational interviews, approximately 25% of partners participated in program events and activities and an additional 50% expressed interest in participating in the future. Some partners (35%) have expressed interest in engaging with participants on a deeper level and are willing to commit to a long-term career mentorship role with one or more individuals in the program.

Connecting partners and participants

The process of connecting program participants with partners in our network begins with trainee self-assessment, which is facilitated by the FUTURE Career Exploration workshop series through assignments and in-class exercises. In early workshops, participants examine and discuss their skills, interests, and values and engage in exercises to determine the current parameters of their career exploration (e.g., career sectors, job categories, timeline, geography, deal-breakers). They then meet one-on-one with a workshop co-instructor to discuss career exploration goals and to determine what knowledge they hope to gain from informational interviews with a network partner. Based on these conversations, the co-instructors write a formal introduction that includes the participant's research specialties, motivations, and goals (participants make final adjustments as necessary). The partner network specialist, a member of our program staff team, uses this information to prepare a curated list of appropriate partners, explaining how each one meets the needs of the participant. The participants select the partners they would like to be introduced to and the network specialist connects them by email. Participants first gain experience with informational interviewing by practicing with groups in the workshop series prior to connecting individually with a partner. Once the informational interview is conducted, trainees share the main takeaways from their conversations with their fellow program members, and both participants and partners are asked to evaluate the quality of the experience. As data collection and analysis is in process, we will publish this information separately.

Partner motivation and satisfaction

When partners were asked what drew them to join our partner network, their answers coalesced around common themes: the desire to provide the kind of career insight they wish had been available to them as trainees, general interest in mentorship, and gratitude for support they experienced in their own training coupled with a wish to return the favor to someone else. After participating in informational interviews and engaging with other aspects of the FUTURE program, partners express satisfaction with their interactions, citing their enjoyment in sharing their career experiences, their excitement when trainees they had mentored land jobs, and their joy in lending participants hope and inspiration for a fulfilling career. An added benefit noted by some partners was the ability to meet and network with other partners.

Participant experience

It is clear from participant feedback that graduate students and postdocs value exposure to careers outside academia. Access to the Ph.D. professionals in our partner network is frequently cited by trainees as one of the most valuable aspects of the FUTURE program. After engaging with network partners, trainees reported gaining greater awareness of career options, exposure to careers they hadn't considered, insight into the day-to-day activities of certain professions, a better understanding of how their skills could be applied in specific job categories or career sectors, and direct guidance on how to prepare for specific career roles. Participants also indicate that engagement with FUTURE partners also yields other, less direct benefits such as a greater understanding of the power of networking and increased comfort with seeking out and soliciting advice from professionals in positions or careers of interest.

Summary

In the FUTURE program, professionals in the partner network provide successful models of Ph.D. scientists who have transitioned from academia to fulfilling careers in all sectors of the workforce. We have witnessed how interactions with these professionals instill confidence in career planning and provide motivation for trainees to prepare for their own career transitions. The FUTURE Partner Network has proved to be a valuable investment for our program and an effective strategy for supporting Ph.D.-trained scientists in their career exploration and preparation for a broad range of careers, including tenure-track faculty positions.

Acknowledgments

This work was supported by the National Institutes of Health Common Fund through grant number 5DP7OD018426 (PIs: Frederick Meyers, MD, Lars Berglund, MD, Ph.D., Andrew Hargadon, Ph.D.) and the National Center for Advancing Translational Sciences, National Institutes of Health, through grant number UL1 TR001860 (PI: Berglund). The content is solely the responsibility of the authors and does not necessarily represent the official views of the NIH.

References

[1] National Institutes of Health (US) Report. Biomedical research workforce working group report. 2012.
[2] FUTURE. Career exploration for graduate students and postdoctoral scholars. February 2019. Available from: www.future.ucdavis.edu.
[3] Mathur A, Brandt P, Chalkley R, Daniel L, Labosky P, Stayart C, Meyers F. Evolution of a functional taxonomy of career pathways for biomedical trainees. Journal of Clinical and Translational Science 2018;2(2):63−5.

Keys to successful implementation of a professional development program: Insights from UC Irvine's GPS-BIOMED

Harinder Singh[a], David A. Fruman[b]

[a]GPS-BIOMED, University of California at Irvine, Irvine, CA, United States; [b]Department of Molecular Biology and Biochemistry; GPS-BIOMED, University of California at Irvine, Irvine, CA, United States

Introduction

In 2014, the University of California, Irvine (UCI), received a five-year award from the NIH-BEST program to fund innovative professional development activities for students and postdoctoral fellows in the biomedical sciences. The name of our program at UCI is Graduate Professional Success in the Biomedical Sciences (GPS-BIOMED) [1]. This program is open to a broad range of trainees across diverse disciplines at our campus in Irvine. Ph.D. students from 10 different Ph.D. programs and postdoctoral fellows from more than 20 departments are eligible to participate. These programs and departments span four different Schools: Biological Sciences, Medicine, Physical Sciences (mainly Chemistry), and Engineering (mainly Biomedical and Chemical Engineering). Participation has increased each year; there are 290 Ph.D. students and 57 postdoctoral fellows in the current year (2018–19), compared to 82 students and 17 postdocs in the first year (2014–15). Participants choose from a wide range of activities that we categorize under four pillars: Explore, Train, Experience, and Transition. Many activities are open to trainees who are not officially registered in the program. Registered trainees accrue additional benefits based on participation, such as access to fee waivers for courses offered through our UCI School of Continuing Education (Fig. 9.1).

In the four and a half years that GPS-BIOMED has been operating, we have experienced many successes and have helped our graduates transition into rewarding careers, but we

BEST: Implementing Career Development Activities for Biomedical Research Trainees
https://doi.org/10.1016/B978-0-12-820759-8.00009-7

FIG. 9.1   UCI GPS-BIOMED logo.

have also encountered obstacles and frustrations. As we contemplate the approaches that have worked best, several key concepts have emerged. The theme of this chapter will be to define a few such "Keys to Success" and illustrate them with examples from our program. We will also highlight some approaches that did not work as well as we envisioned, and how we overcame those challenges. We hope that this information will be useful to other universities and research institutes that wish to build or enhance their professional development activities for trainees in the biomedical sciences.

Keys to Success

Ph.D. graduates in biomedical research fields have many options available for fulfilling careers. While many still seek academic faculty positions, others are interested in different jobs within the academic research enterprise or in various careers in the public and private sectors. One key to helping trainees achieve these goals is to *leverage existing resources*. At most universities and research institutes, all stakeholders (trainees, faculty, and administrators) recognize the importance of preparing trainees for diverse careers. Indeed, over the last decade, most institutions have established and expanded professional development activities for Ph.D. students and postdocs. Often, these are organized within campus units of limited scope, such as departments, training grant programs, and postdoctoral associations; UCI has many such small-unit professional development activities. The administrator of a centralized program can establish collaborations with multiple units across campus to support their efforts, share best practices, improve publicity, and broaden participation. Our GPS-BIOMED program, thus, frequently co-sponsors such activities, helping advertise through a central web portal, email list, and social media platforms. When possible, we contribute funds toward the costs of speakers, food, etc.

Another way to leverage existing resources is to establish collaborations with campus units that can provide unique programming or experiential learning opportunities. For example, technology transfer offices are a good place for trainees to conduct part-time internships to learn about intellectual property and licensing. These offices will sometimes give preference to applicants who are highly motivated and have completed prior professional development training. Five GPS-BIOMED trainees have completed internships at UCI's Invention Transfer Group and three have transitioned into careers in this field (Fig. 9.2).

It is also crucial to *develop new partnerships that add high value* to the overall training program. A relevant example from GPS-BIOMED concerns our efforts to improve training in science communication. Like most other NIH BEST awardees, we identified an urgent need to

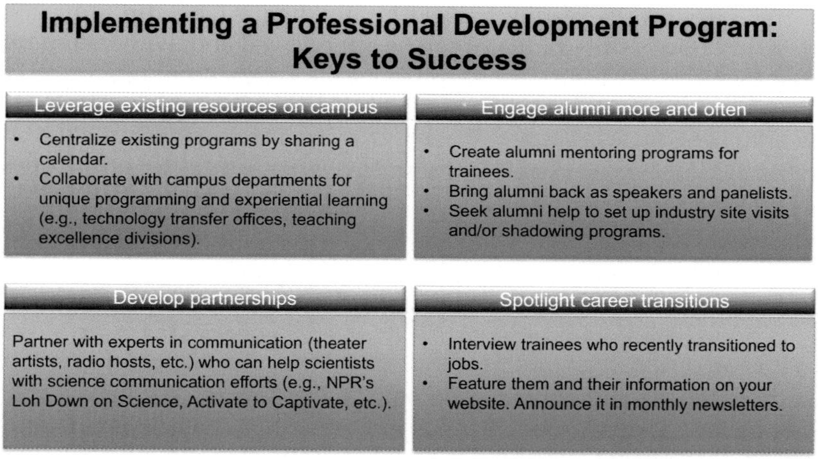

Implementing a Professional Development Program: Keys to Success

Leverage existing resources on campus	Engage alumni more and often
• Centralize existing programs by sharing a calendar. • Collaborate with campus departments for unique programming and experiential learning (e.g., technology transfer offices, teaching excellence divisions).	• Create alumni mentoring programs for trainees. • Bring alumni back as speakers and panelists. • Seek alumni help to set up industry site visits and/or shadowing programs.
Develop partnerships	Spotlight career transitions
Partner with experts in communication (theater artists, radio hosts, etc.) who can help scientists with science communication efforts (e.g., NPR's Loh Down on Science, Activate to Captivate, etc.).	• Interview trainees who recently transitioned to jobs. • Feature them and their information on your website. Announce it in monthly newsletters.

FIG. 9.2 Sampling of actions that have been key to the success of UCI's GPS-BIOMED.

expand our trainees' oral presentation experience beyond typical research-in-progress seminars and journal clubs. However, most research faculty do not have the expertise to train students in communicating to diverse audiences. We enlisted two communications experts who provided workshops on complementary aspects of presentation skills. The first expert is Bri McWhorter [2], a former actress who obtained an MFA in drama from UCI; she coaches students and postdocs in performance skills and leads our "Elevator Pitch" competition. The second is Sandra Tsing Loh [3], a nationally recognized science communicator and host for the syndicated podcast "Loh Down on Science," who has an adjunct professorship at UCI. She teaches a two-unit course in which students learn how to prepare and deliver a five-minute TED-style talk about their research. Both of these partnerships have received very high evaluation scores by participants. While not every campus has nationally known science communication instructors, the lesson is that the best place to find the right instructor is often outside the science departments themselves. Other types of university partnerships could include faculty or alumni from schools of business, law, or other.

A third key to success is to *engage alumni early and often*. Ph.D. graduates and former postdocs are an underutilized resource. These professionals have previously navigated career transitions and have relevant experience and connections in different workforce sectors. Importantly, we have found most of our alumni to be highly receptive to donating their time to give back to the university. Most frequently, this occurs through campus visits, where alumni give career talks or serve on career panels. It can also take the form of speed-mentoring sessions or individualized mentor pairings. Alumni also are ideal for hosting informational interviews and for arranging group visits to biotech/pharma companies. Ph.D. and postdoctoral alumni can also help bring to campus other non-alumni who work locally and are interested in connecting with (and recruiting) trainees.

Although there is increasing support for the concept of graduate professional development, it always helps to remind all stakeholders about positive outcomes. Therefore, a fourth Key to Success and sustainability is to *spotlight successful career transitions* [4]. Career progress of program graduates can be readily tracked by building a LinkedIn group of current and former participants. Successful graduates can be interviewed and, with their permission, their interviews can be recorded and transcribed. In addition, recent alumni often write the program leaders to express gratitude for help with a career transition. These success stories should be highlighted in newsletters, the program website, and marketing materials. We have an alumni spotlights page on our website [4] where we include helpful quotes from program alumni. Samples include:

- "One thing I like about GPS-BIOMED is that they do a lot of these events that get you in contact with people that are out in the field, in various fields. Without contacts it is very, very hard to get your foot in the door."
- "The Regulatory Affairs course in particular was very beneficial for me to have on my CV or resume because it showed employers that I was committed to expanding my skill set."
- "I learned everything I know in terms of soft skills in industry from the SciPhD workshop. Being able to communicate, being able to interview well, being able to write an effective resume and also a cover letter are really important. These skills are pretty much essential in order to get a job in industry."

Universal skills to enhance training for diverse career paths, *including academia*

A good way to ensure buy-in from faculty and administration is to emphasize how the program develops skills that are essential for all Ph.D. graduates, including those intent on pursuing a career as an academic research leader. Below are examples of four skill sets that are not always developed through the course of traditional Ph.D. training, but that are certain to be beneficial for trainees preparing for academic careers.

a. Science communication to diverse audiences: During graduate school and postdoctoral training, there are many opportunities to hone one's skills presenting a detailed research seminar to experts in the field. But most academic departments do not prepare trainees to present their science in an engaging way to other audiences. During the course of a professor's academic career, there are many occasions where it is essential to make a pitch to non-experts. Examples include talking to philanthropists, science reporters, and venture capitalists. Strong communication skills make those interactions more successful; moreover, explaining science without jargon sure helps when trying to transmit the importance of our work to family members and friends!

b. Business basics (including technology transfer): In today's scientific enterprise, many academic researchers collaborate extensively with biotech and pharma companies through sponsored research and/or advisory positions. Some faculty develop their own intellectual property to establish start-up companies. Familiarity with the business of science will make such endeavors more successful and bring new revenue streams to the university.

c. Project management: It is true that Ph.D. students and postdocs learn to manage multiple projects over the course of their training. However, project management skills are largely gained by trial and error, with variable guidance from faculty mentors. A strong understanding of project management makes the transition to a faculty position more successful. Courses and certificate programs in project management can help build these skills for academics and for those moving to research positions in industry.

d. Working in teams: Science is a collaborative endeavor, and few of us publish single-author or two-author papers anymore. Most students and postdocs do gain experience working in collaborative teams. Yet, as with project management, the skills needed for successful team science are often learned through experience. There is a stereotype that academic leaders don't always "play well in the sandbox." Workshops and mini-courses on team science introduce key concepts that increase efficiency, prevent misunderstandings, and improve team harmony.

There is no reason why a professional development program cannot incorporate activities that are designed specifically for trainees on the academic track. For example, a program can deliver workshops in grant and fellowship writing or in chalk talk preparation.

Examples relevant to seven career pathways

In this section, we list seven common career pathways and discuss how GPS-BIOMED has followed the abovementioned Keys to Success. For the sake of brevity, these summaries do not list specific alumni spotlights. As noted above, our program website has a page that spotlights these career transitions and provides guidance to current trainees from the perspective of recent graduates.

Biotechnology and pharmaceutical industry

Careers in biotech/pharma have always been an attractive destination for biomedical trainees. Therefore, most current faculty have several former lab members working in biotech/pharma. We *leveraged these existing resources* by reaching out to UCI faculty, and asking them to send names of lab alumni who might be interested in participating in GPS-BIOMED programming. Since most faculty maintain a record of former trainees for their own academic file and for training grants, it is a simple task for them to send a few names and contact information. These contacts allowed us to *engage alumni early and often*. Such alumni return to campus for career panels and "Life Beyond the Ph.D." seminars. Some alumni participate in our Alumni Mentor Program, and others invite groups of trainees for site visits to their companies.

Another existing resource we have leveraged is our School of Continuing Education [5]. The Dean of Continuing Education at UCI grants fee waivers and discounts to GPS-BIOMED trainees who have completed at least five hours of professional development activities. Popular courses relevant to careers in biotech/pharma include Regulatory Affairs (mentioned in a trainee quote above) and Project Management. Many of these courses are given online so they can be completed without disrupting research time. Although not all institutions can negotiate fee waivers for extension courses, they could still inform trainees about continuing education courses that are available to them.

Business and entrepreneurship

At the time of this writing, the US economy has been expanding for a decade, business taxes have been cut, and the biotechnology industry is booming. There are countless new startups backed by a deep pool of venture capital. More and more of our trainees are attracted to the startup space and many prefer such destinations to large pharma. How are we helping them explore and prepare for such careers? For this, we have *leveraged an existing resource*, UCI Applied Innovation [6]. This office is a hub for inventors, investors, and the local business community. Our trainees have participated in business plan competitions and entrepreneurship boot camps. We are currently working with Applied Innovation to develop internships where our trainees can assist with market research and/or write SBIR grants. We have also *engaged UCI alumni* who are entrepreneurs and have encouraged them to participate in our Alumni Mentoring Program.

In addition, we have *established a new partnership that adds value*. Specifically, our program has hosted "Business of Science" boot camps offered by SciPhD. As noted on their website [7], "This comprehensive 36-hour program makes academic scientists "business-ready" for professional positions, whether in academia, government, or industry. SciPhD's experiential "learn by doing" methodology provides hands-on training for students to experience the essential skills valued and required by professional organizations." Participants have given high ratings to SciPhD boot camps, as noted in the testimonial quoted earlier in this chapter. While the cost of this external contractor was readily absorbed by our NIH BEST grant, other funding models can be envisioned, including partnering with other local institutions or charging participants a fee.

Data science

There is a growing market for scientists trained in database management, "Big Data" analysis, machine learning, and other aspects of "Data Science." Several of our trainees have taken Continuing Education courses in data science, thus *leveraging an existing resource*. In addition, our program has helped to publicize the UCI Data Science Initiative [8], thereby informing trainees about valuable training activities, such as short courses and certificate programs.

Intellectual property

As the pace of biomedical innovation has grown, so has the need for Ph.D.-level scientists to apply their expertise toward technology transfer and patent law. To this end, we have *leveraged an existing resource*, the Invention Transfer Group within UCI Applied Innovation, to establish a part-time internship program where trainees gain hands-on experience in technology market research, the patent application process, and licensing. This has been our most popular internship program, with five trainees participating over three years. We have also *engaged a UCI postdoctoral alumna*, a Senior Director of Licensing in the Invention Transfer Group, to participate in our Alumni Mentor Program and serve on the GPS-BIOMED Stakeholder Committee, an internal advisory board. We have also established a *new partnership that adds value* by establishing an internship program with a local patent law firm.

Science communication

While all Ph.D. trainees strive to develop strong communication skills for diverse careers, some find that science communication itself is an attractive career path. Biotech and pharma employ Ph.D. graduates in various communications roles, including publications management, medical science liaisons, and marketing. The Keys to Success section above provided details of how we *added valuable new partnerships* with communications experts Bri McWhorter and Sandra Tsing Loh. We have also *engaged alumni* who have established careers in medical writing and communication; one of them served as an alumni mentor to two of our trainees in the summer of 2018.

Growing interest in science communication led a group of our students to form a new Meetup group known as "Brews and Brains" [9]. This organization organizes local events where our researchers meet with community members over beer and snacks, and present 5- to 10-minute research talks at a level understandable to professionals in science and engineering fields.

Science policy

The last few years have seen increasing activism among scientists. From the "March for Science" marches and rallies, to the public policy fellowship programs offered by AAAS and other professional societies, there are more and more opportunities to get involved in science policy and advocacy. Some trainees are interested in careers in science policy, whereas others envision part-time or volunteer roles in advocacy.

To serve the needs of this growing constituency, we have *added valuable new partnerships* by reaching out to local non-profit agencies. These include the Orange County Alzheimer's Association and our local chapter of the American Cancer Society Cancer Action Network. These agencies have offered opportunities for trainees to participate in writing policy statements aligned with their mission. With letters of support from these partners and UCI campus offices, we were successful in obtaining a grant from the Burroughs Wellcome Foundation to support the Public Policy Prep (P3) program for trainees in the biomedical sciences, which we developed by *leveraging existing resources*.

One of the resources we utilized for the P3 program is a self-organized Science Policy Group of Ph.D. students and postdocs. Members of this group have helped guide our choice of activities for P3. Another campus resource we have tapped is the Office of Community and Government Relations. Staff members in this office have advised us on how to arrange for trainees to engage with legislative offices and how to plan for visits to our state capitol and Washington, DC. Lastly, we have engaged with our Department of Public Policy to identify a faculty member willing to lead a workshop with biomedical trainees.

Teaching

An attractive academic career pathway for some Ph.D. trainees is to obtain a faculty position at a teaching-intensive university or college. In fact, one of our best-attended career panels focused on such careers. To assemble the panel of speakers, we *engaged UCI alumni* who teach at California State University, Long Beach, and Chapman University. However, attending a career panel is only the first step in preparing for a teaching career. To be competitive for teaching-intensive faculty positions, and to meet expectations if hired, requires specific pedagogical training and experience as an instructor of record. The teaching experience that most Ph.D. students acquire is limited to grading exams and running discussion sections.

To meet demand for better preparation for teaching careers, we *leveraged existing resources*. At UC Irvine, each department in the School of Biological Sciences has two or three faculty known as Professors of Teaching. These faculty conduct pedagogical research and have a heavier teaching load than research faculty, mostly in large, lower division courses. Collaborating with these teaching faculty, we developed a Teaching Apprenticeship in STEM (TAP-STEM) program. Each fall, we select Ph.D. students and postdocs who have had ample prior experience as TAs and an interest in teaching careers. These trainees attend workshops given by the UCI Division of Teaching Excellence and Innovation; they also receive year-long mentorship by Professors of Teaching, focused on one of the large undergraduate biology core classes. The following summer, they are eligible to teach the summer session as instructor of record.

Challenges

So far, this chapter has defined Keys to Success and given examples of how UC Irvine applied them to develop successful programming. But like all NIH BEST programs, we tested ideas that did not work as well. Here, we will briefly mention two of the less successful efforts.

Experiential learning (internships, job shadowing) can be a key element of career exploration and skill development. However, Ph.D. students and postdocs must balance their interest in such activities with the need to complete their research projects in competitive fields. Moreover, there are fewer existing internship programs for Ph.D. students and postdocs as there are for undergraduate and master's level students. When we applied for NIH BEST funding, we had obtained written commitments from several local companies to establish internship programs for UCI biomedical Ph.D. students and postdocs. After the grant was awarded, few of those plans came to fruition. Companies cited other priorities or turnover in leadership, or just did not respond to follow up. Initially, our staff devoted considerable time to these efforts with minimal success. Although the internships that we did establish were successful (as reported by trainees and hosts), the personnel effort per trainee was not sustainable. We eventually moved to a model with two elements that limit effort by program staff: (1) the program curates and advertises existing internship programs to our trainees, and (2) trainees can develop new internships through their own connections if they are interested in so doing. When trainees do the legwork to identify an internship host, staff can help with logistical aspects, such as workman's compensation and confidentiality agreements. Of note, the most consistent success that we have had was with part-time internships in campus offices — like the Invention Transfer Group described above. As other NIH BEST programs had more success establishing internship programs, we refer readers to those other chapters for more details.

A second challenge that we encountered was in preparing trainees for networking events. One of our early partnerships was with a local technology industry group. They agreed to include groups of our trainees at networking events that are mainly attended by established professionals in the biotech or medical device fields. Without formal preparation, our trainees felt out of place and did not know how to approach the experienced professionals. Later on, we developed workshops on how to prepare for networking events and individualized mentoring opportunities. These workshops were instrumental in improving the outcomes of later activities such as our Alumni Mentor Program.

Concluding thoughts

All biomedical research institutions aspire to nurture the career goals of their trainees. However, with limited financial resources, it is essential to make cost-effective choices. The specific examples of successful activities at UC Irvine might not be feasible elsewhere. Each institution has its own strengths that can be leveraged to build a valuable program. Even modest enhancements can improve the campus culture and help make trainees feel more supported [10].

During program building and evolution, it is important to engage the trainees for "bottom-up" decision making. Although formal assessments of the program and activities are helpful, extensive surveys are burdensome and response rates vary. Our students and postdocs have self-organized a trainee "think tank," which meets every quarter to provides valuable feedback to leadership, and sometimes seeds the development of new activities. Thus, program development should be a collaborative effort involving eager trainees, engaged faculty, and supportive administrators.

Acknowledgments

The authors would like thank the NIH BEST initiative for awarding us grant DP7-OD020321 to establish the BEST program at UCI. The current GPS-BIOMED leadership would like to thank previous Associate Director Emma Flores-Kim, Ph.D., for building and expanding the program from 2014 to 2018 and for making it a unique professional development program in the Southern California region. Additionally, we would like to thank Dean Frances Leslie, Ph.D., and the UCI Graduate Division; Dean Gary Matkin, Ph.D., and the Division of Continuing Education; and all of our advisory and stakeholder committee members for their continued support and guidance over the years.

References

[1] Graduate professional success in the biomedical sciences (GPS-BIOMED). Available from: https://gps.bio.uci.edu.
[2] Bri McWhorter. Available from: https://www.activatetocaptivate.com.
[3] The Loh down on science. Available from: https://www.npr.org/podcasts/381444663/the-loh-down-on-science.
[4] Alumni Spotlights. Available from: https://gps.bio.uci.edu/alumni-spotlights/.
[5] University of California, Division of Continuing Education. Available from: Extension school, https://ce.uci.edu.
[6] Applied Innovation. Available from: http://innovation.uci.edu.
[7] Business of science for scientists. Available from: http://sciPhDcom.
[8] University of California, Data science initiative. Available from: http://datascience.uci.edu/about/.
[9] Brews and Brains. Available from: https://www.meetup.com/Brews-and-Brains/.
[10] Building a Career Development Program for Biomedical Trainees. Open online course available at: https://uciunex.instructure.com/courses/10505.

The Motivating INformed Decisions (MINDs) Program — The University of California San Francisco BEST program for career exploration

Gabriela C. Monsalve, Jennie Dorman, Bill Lindstaedt, Elizabeth A. Silva, Keith R. Yamamoto, Theresa C. O'Brien

University of California, San Francisco, CA, United States

Overview of the MIND program

The vast majority of people earning biomedical Ph.D.s find work, but the number of Ph.D.s who obtain tenure-track faculty positions now represents a shrinking minority. The University of California San Francisco (UCSF) MIND program [1], funded by an NIH Broadening Experiences in Scientific Training (BEST) award, was an experiment in career development and exploration. We aimed to provide better resources for students and postdocs who wished to explore the wide range of possible career outcomes that are available to biomedical trainees and better support for the faculty who mentor them.

The most immediate goal of the MIND program was to create, deliver, and test the effectiveness of a comprehensive career development intervention for early-stage Ph.D. students and postdocs and their mentors to address some of the specific knowledge gaps and mismatched motives that can derail the career decision-making process. The long-term goal was to change the culture at UCSF in a fundamental way, cultivating a community in which students and postdocs felt supported in their exploration of the broad range of career options in the biomedical workforce.

BEST: Implementing Career Development Activities for Biomedical Research Trainees
https://doi.org/10.1016/B978-0-12-820759-8.00010-3

Major aims of the MIND experiment

The UCSF BEST grant was organized around three important initiatives. First, providing coursework for students and postdocs to develop the skills and tools they need to explore a variety of career paths. Second, connecting these trainees with MIND program partners to take a closer look at different careers (MIND partners are biomedical Ph.D.s in careers both within and outside academic research and will be discussed in detail below). Third, assessing the types of support and information that faculty need to mentor trainees whose career interests diverge from the academic path, and building resources to support these endeavors.

The MIND program: Career exploration for Ph.D. students and postdocs

The MIND program [2], which serves around 80 participants annually, is an eight-month career development and exploration program specifically developed for UCSF graduate students and postdocs in the basic life sciences. Acceptance into the two-phase MIND program follows a brief application, the main purpose of which is to eliminate participants who misunderstand the purpose of a career *exploration* program; applicants seeking help with getting a job right away are referred to other programs at UCSF that are more suited to that goal.

The first phase of the MIND program, the MIND Catalytic Course [3], is a targeted course informed by career development literature and designed to teach career exploration skills to biomedical scholars. In this class setting, trainees develop the basic skills needed to investigate career options that are best suited to their skills, interests, and values. These full-day class sessions occur during three Saturdays between October and December [4].

The MIND Catalytic Course is career-neutral and encourages participants to explore whatever careers are of greatest interest to them regardless of whether they are inside or outside of academia. It is designed to serve as a springboard that primes trainees to fully exploit the second part of the MIND program, in which they direct their own career exploration activities with the help of a team of peer coaches (more information below). The course is designed to be *catalytic*: it provides scholars with the *substrates* they need — the tools and conceptual frameworks required for career exploration — to activate their career exploration journey and *catalyzes* a career planning *reaction* that takes place during the course of the MIND program. As a requisite for entry into the course, participants agree to attend all three sessions and complete all of the homework.

By the end of the MIND Catalytic Course, Phase 1 of the MIND program, trainees are able to:

1. List the activities leading to successful career transition and reinvention.
2. Identify two to three careers that best fit their unique skills, interests, and values.
3. Identify professionals in their careers of interest whom they would like to add to their network.

4. Define their goals for an informational interview and prepare the materials needed to initiate and conduct one.
5. Synthesize and reflect on the information they collect about careers of interest to assess alignment with their skills, interest, and values.

After the MIND Catalytic Course, trainees enter the second phase of the MIND program, meeting twice per month, January to May, in peer teams [5] to explore various career options. Phase 2 of the MIND program, the peer team phase, is designed to continue to support and engage participants through the inevitable ups and downs of career exploration [6]. Each peer team comprises a mix of five to seven postdocs and graduate students *exploring a variety of career paths*. A unique component of the peer teams is that they are not organized by career interest. This organization is intentional, as we know that participants often explore many careers simultaneously and that their career interests may change during the course of their exploration. Thus, the added benefit of a diverse peer team is the multiple perspectives that each individual brings, helping all team members grow and diversify their professional networks. The peers in the team challenge and support each other as they structure their time, brainstorm collectively how to overcome obstacles, meet their deadlines, and find confidence in their career directions.

Each team is assigned two MIND facilitators. These trained facilitators — primarily staff or students who are alumni of the MIND program — help structure and facilitate conversations among the team members to keep the focus on problem solving and to ensure that everyone receives attention from the group. Peer teams meet every other week, during which time each individual participant sets career exploration goals, and together the team members hold one another accountable to those goals.

The principal tool for effective career exploration is the informational interview, which is taught to MIND participants in the first phase of the program. MIND participants connect with professionals and UCSF alumni [7] in a range of fields and jobs to gather reliable and comprehensive information about their options. Trainees share this knowledge and experience with their peer teams and document it in a knowledge base, which serves as a resource for all current and past MIND participants.

By the end of Phase 2 of the MIND program, the peer teams phase, trainees:

1. Prepare and conduct three or more informational interviews with professionals in their career(s) of interest.
2. Break down their career exploration journey into achievable goals.
3. Practice and improve their teamwork and communication skills.
4. Evaluate whether or not the career(s) they are considering align with their skills, interests, and values.
5. Develop a career plan that helps boost their career confidence.

To ensure that each participant and team continues to make progress, MIND convenes for two additional, all-cohort meetings. These (roughly) five-hour sessions provide supplementary strategies and tools (including the MIND Career Exploration Road Map [8]) to recharge participants' career exploration journeys.

The MIND program ends with an all-cohort "finale" gathering during which the MIND team celebrates the progress of all program participants.

MIND bank: A repository of Ph.D.-trained professionals for MIND participants

The MIND program emphasizes informational interviews [9] as the key strategy for exploring careers effectively. To accelerate and support MIND participants seeking to meet with professionals, the MIND program created MINDbank, a crowd-sourced repository of biomedical Ph.D.-trained professionals (MIND partners) pursuing a variety of different careers both within and outside academia.

Prior to entry into the MIND program, many Ph.D. student and postdoc participants do not have *any* experience with requesting an informational interview. Given this anxiety-provoking "obstacle," MINDbank was designed to lower the initial barrier to the informational interview process by providing quality, appropriate, and enthusiastic career professionals who *wanted to speak with UCSF trainees*. MIND staff initially used their own professional networks to recruit partners and then expanded their recruitment efforts via existing partnerships with UCSF and MIND alumni and the California Life Science Association [10].

Potential MIND program partners were initially contacted via email and were given a link to the MIND program's partner webpage, where they found a promotional video from Keith Yamamoto, one of the co-PIs of MIND and a long-standing faculty leader. The video encouraged their participation, starting with the completion of an initial (but very brief) online interest form. The MIND program director then conducted a 10- to 15-minute phone interview with every individual who completed the interest form. While labor- and time-intensive, these additional steps ensured that information within the database was up-to-date and accurate, and helped to establish a positive and personal relationship with the partners.

After the phone interview, MIND partners were asked to complete a brief assessment of the skills and tasks required for their current position. The MIND program originally planned to anonymize and disseminate aggregated data by career type as a public resource for other Ph.D. students and postdocs interested in gaining more information about specific skills and tasks associated with particular career trajectories. On the programmatic level, all of the data collected would be used to assist trainees in finding partners who matched their career interests and to facilitate the initial communication leading up to an informational interview. A sophisticated cloud-based software solution called MINDbank was built to support these functions.

By 2015, after collecting 190 unique reports on the skills and tasks associated with the careers of the 258 MIND partners we had at the time, our home institution adopted a more established alumni networking platform through Graduway. UCSF Connect [11], a version of the Graduway platform customized for our campus that enables matching between professionals and trainees for the purpose of career and professional development, held a greater promise for sustainability and, despite lacking some of the unique features we had envisioned, swayed us to ultimately abandon MINDbank. Now, instead of recruiting partners through the MIND program, the Office of Career and Professional Development broadly recruits alumni for participation in UCSF Connect, and MIND program partners access informational interview contacts directly through this platform. Original MIND partners were invited to join the UCSF Connect platform in order to stay connected to the MIND program.

Beyond facilitating informational interviews, the other major goal for MINDbank was to serve as a repository of "Trainee Takeaways," synopsis reports created and submitted by MIND participants after completing informational interviews with MIND partners representing a wide variety of career fields. While the MINDbank platform was being developed, trainees were asked to submit their Takeaways via the MIND program internal website (a virtual learning platform offered through the UCSF library). In this manner, the MIND program collected a total of 144 Trainee Takeaways during the NIH-funded portion of the MIND program, each outlining a trainee's observations gleaned from their one-on-one interactions with professionals. We ultimately decided to leave all Trainee Takeaways from previous years on the MIND program internal website and made them available to current and former MIND participants. The library of Trainee Takeaways continues to grow, as each new cohort is encouraged to read the Takeaways related to their careers of interest prior to conducting their own informational interviews and adding to the library themselves. The purpose of this activity is to provide trainees with upfront knowledge about a career field so that they can bring up questions that "dig deeper" during their valuable one-on-one informational interview time. The Takeaways library also leverages the time of MIND partners and other UCSF Connect members who conduct informational interviews with MIND participants, allowing their insights to be shared with multiple trainees.

UCSF faculty support via needs assessments and resource development

The third aim of the MIND program was to change faculty attitudes in a fundamental way so that UCSF mentors would be more accepting and supportive of a wide variety of trainee career outcomes. Specifically, the MIND program's interactions with faculty aimed to:

- Provide them with greater knowledge of career path alternatives for their trainees.
- Alleviate concerns about trainees pursuing careers other than academic research.
- Facilitate greater support for mentoring responsibilities from peers and the institution.
- Enable the adoption of higher levels of innovative mentoring practices.

To accomplish these goals, the MIND program team, comprising both faculty and staff, worked with UCSF faculty to understand their perspectives and concerns about current training practices, identify and fill specific knowledge gaps, and develop resources to support faculty mentoring efforts.

Over the course of four years, the MIND team convened multiple faculty working groups; conducted in-depth, one-on-one interviews with 13 faculty who represented a wide range of views on the issue of trainee career outcomes; and offered presentations about the MIND program (and BEST efforts in general) during faculty departmental meetings. As the MIND program spoke with more and more faculty, interview data revealed opinions that were generally more supportive than the interviewers expected. The faculty were also more supportive of non-academic career options than students perceived them to be.

To that end, the MIND program worked with faculty to develop a system that could help support them in their mentoring of UCSF trainees. This innovation, called the "Career Outcomes Transparency Pilot," [12] has the vision of providing lab alumni career outcomes information on faculty members' profile pages. The initial goal was to partner with one basic science department to explore feasibility, develop a working prototype and collect feedback from faculty. To accomplish this, the MIND team created a list of every former trainee from the department members' laboratories, and then collected career outcomes (initial job and current position) information from as many former trainees as possible.

The team created lab-by-lab reports and partnered with UCSF's Clinical & Translational Science Institute [13] to create a widget that allowed the career outcomes data to be linked to individual UCSF faculty profiles pages [14]. For this pilot effort, the lab-by-lab trainee outcomes were not available to the public. The MIND team plans to assess the pilot by interviewing individual faculty members as they are presented with the working prototype. We want to understand how the faculty feel about the usefulness of the information being made available in this format, both for their own use and that of current and prospective trainees, and whether they would be willing to have the information made public. Provided the initial group of faculty are supportive and we obtain institutional support, we will consider options for expanding the pilot to other UCSF departments. We feel efforts such as these will have a positive impact on the culture of UCSF, making it more supportive of career exploration and diverse career options.

Experimental outcomes

What worked?

MIND program learning objectives and goals

The intended outcomes of the MIND program were largely realized. Based on preliminary internal measures:

- 96% of MIND trainees reported an improved or a greatly improved ability to articulate their career plans with peers and colleagues.
- 71% of MIND trainees reported a clear sense of the subsequent training required for their career of interest.
- Peer teams provide an effective vehicle for improving participant accountability and motivation for persisting with and planning career exploration activities.
- 77% of MIND trainees report improved satisfaction with their overall training experience at UCSF.

Expansion and institutionalization of the MIND program

One overarching goal of the MIND program was to create scalable programming that could reach a significant proportion of UCSF's biomedical graduate student and postdoc populations. Although the institution already had a number of successful career and professional development programs for this training community, they could generally accommodate only 30 or so trainees at a time — a fraction of the nearly 650 graduate students and over 1,100 postdocs at UCSF. Thus, the MIND program cohorts were designed to grow over time, reaching a threshold level that we hoped met demand and was not impeded by operational constraints like classroom size and available staff.

During the grant period, we successfully expanded the total number of participants served in the MIND course from 32 in year 1 to 92 in year 4. We consistently increased the number of peer teams without increasing the number of trainees or decreasing the number of facilitators per team. During our fourth (and largest) year, the cohort was divided into 13 teams of 6–8 members, for which the MIND program successfully recruited and trained 26 facilitators, 2 for each team. The peer team facilitators were graduate students, postdocs, MIND staff, and other Ph.D.-level staff from other UCSF units, including the Research Development Office, the Office of Technology Transfer, and the Research Management Office.

To recruit so many volunteer facilitators — each of whom must agree to undergo 6–10 hours of training meetings, 20–30 hours of peer team meetings, and however much time is required to organize and support team members — the MIND staff relied heavily on MIND alumni, current students who had already matriculated through the program. Because the MIND program will continue to generate more alumni, we predict that staff will still be able to recruit facilitators in the future who are willing to give back in this way. After several years of experience training new facilitators and defining best practices for successful peer teams, the MIND team published a document called "Peer Mentoring Teams for Career Exploration: Design Considerations and Facilitator Training Manual," [15] which is freely available to anyone interested in setting up their own peer teams.

Based on increasing participation in the MIND trainee program — despite consuming three full Saturdays and two hours of biweekly peer team meetings for six months — the MIND program has built an outstanding reputation on campus. In fact, responses from prospective applicants to the MIND program indicated that approximately half of the cohort applied to MIND based on a recommendation from previous participants. To that end, the MIND program has been fully institutionalized into the Office of Career and Professional Development at UCSF, and will continue to offer career guidance and support for future UCSF graduate students and postdocs.

Leveraging trainees' informational interviews

Another area of accomplishment was the usefulness of "Trainee Takeaways," the synopsis reports submitted by MIND cohort members after they complete an informational interview (mentioned above). All Trainee Takeaways from previous years are available via the MIND program internal website, and current cohort members are encouraged to read the Takeaways related to their careers of interest prior to conducting an informational interview. Seventy 2016–17 cohort members responded to evaluation items concerning Trainee Takeaways, including 70% who had read at least one Takeaway prior to going on an informational interview and 98% who reported the resource to be "valuable" or "very valuable."

Although we decided not to continue developing the MINDbank platform, MIND partner contact information is still available to members of the MIND staff through the unfinished MINDbank platform. MIND staff have continued to promote partners as a resource for MIND participants to identify referrals for informational interview contacts and can provide MIND partner referrals upon the request of a MIND cohort member. Approximately 30% of this year's cohort (2018–19) requested MINDbank partner contacts, and 100% of them reported that they were provided with a contact who met their needs and was helpful. Additionally, MIND participants are encouraged to use UCSF Connect as a referral resource for informational interview contacts.

Job simulation exercises

MIND program activities within the Catalytic Course and the peer teams phases place an emphasis on teaching and on motivating each participant to conduct effective informational interviews in career fields that interest them. However, informational interviews do not provide firsthand experience of the tasks associated with a particular career. Therefore, the MIND program aimed to encourage participants to try tasks typically associated with a career of interest through the use of a job simulation exercise.

A job simulation exercise provides MIND participants with the opportunity to "try out" a career path by completing a task that is emblematic of that career path. For example, an exercise simulating the career path of a professional in the business of science might require the student to complete a competitive analysis of two different drugs on the market. Such a simulation exercise would include instructions, background reading materials, references, and questions for the participant to evaluate their performance on the completed project. Job simulation exercises proved to be much more challenging and time consuming to create than anticipated when we started the MIND program, so we launched a separate effort to create an entire library of simulations. Although this separate effort was not technically funded by MIND or the BEST program, the original idea and early prototypes of job simulations were generated by the MIND PIs; developed into a separate, fundable project using former MIND staff; and supported by former MIND program trainees, who have been outstanding volunteers in creating and testing the simulation exercises. The job simulations were ready to roll out for use by trainees during our final year of the MIND program.

Thi Nguyen, a staff member within UCSF's Office of Career and Professional Development, and Liz Silva, (former) MIND program director, served as co-PIs on a Burroughs Welcome Fund Career Guidance for Trainees proposal, which was granted for 2016—17. Using these additional funds, Nguyen and Silva developed a process for creating job simulations; recruited volunteer alumni and trainees to solicit, write, test, and publish the exercises; and began fleshing them out. As of this writing, more than two dozen job simulations have been created covering fields such as intellectual property, research administration, medical writing, freelance science journalism, science editing, science policy, and business development.

What didn't work?

MINDbank's primary goals were met with alternate institutional tools

The primary goals of MINDbank were to facilitate matching trainees with Ph.D. professionals for informational interviews and to serve as a publicly accessible repository of crowd-sourced career information. As discussed above, we ultimately decided to move to institutionally-supported tools that serve the wider UCSF community, facilitating the matching of trainees to Ph.D. professionals through UCSF Connect [11], and using a collaborative learning environment platform provided by our library for sharing "Trainee Takeaways." The main reason for moving away from the MINDbank platform was sustainability; however, the decision meant that we were not able to make the career information available to those outside of UCSF since the tools we are leveraging are open exclusively to UCSF trainees, faculty, and staff. That said, we still believe there is value in pursuing a resource like MIND-bank that can serve the broader biomedical training community and look forward to tackling this problem in the future.

MIND beyond the BEST grant

Institutionalization of the MIND program

The MIND program has been institutionalized at UCSF with support from Elizabeth Watkins, the dean of the Graduate Division. The dean was one of the key stakeholders in the original grant application, providing feedback on early concepts and a formal letter of support. Early during the grant application period and again during the funded period, discussions between Dean Watkins and the MIND team touched on the long-term future of the MIND program and on the perception among campus leadership that the MIND team would be testing a unique pilot program, that, if successful, would warrant permanent funding at the close of the grant period. One of the MIND PIs reports directly to the dean, so UCSF's BEST program was fortunate to have important relationships in place early on — even before receiving the grant award.

The MIND program is one of many career and professional development co-curricular offerings at UCSF. One of the benefits of its institutionalization is that the MIND program's new home is within the UCSF Office of Career and Professional Development (OCPD). OCPD has decades of experience preparing students and postdocs for their professional lives after UCSF, so the program's objectives are strengthened by its co-location within OCPD and by the additional expertise provided by its staff.

BEST provided a platform from which we were able to design, pilot, test, optimize, and ultimately institutionalize a program model for guiding large numbers of our Ph.D. students and postdocs through the career exploration process. Additionally, the fact that our MIND program had NIH funding imbued our efforts with the credibility necessary to open productive conversations with faculty and to pursue a real change in the culture. We are confident that, overall, our faculty and campus leadership have become more supportive of our trainees' career exploration efforts as a result of the BEST program experiment.

Acknowledgments

This work would not have been possible without funding from the National Institutes of Health through an NIH BEST DP7 Common Fund award (5DP7OD018420). We would like to thank our MIND trainees, external partners and alumni, advisory committee members, and the leadership at UCSF for their advice, support and enthusiastic participation. We could not have created such a successful program without you, and your contributions will endure through MIND programming at UCSF. We are also grateful for our colleagues in the BEST Program Consortium for sharing best practices and lessons learned.

References

[1] MIND: motivating informed decisions. Available from: https://mind.ucsf.edu/. [31 July 2019].

[2] O'Brien, TC, Yamamoto, KR, Lindstaedt, B, Dorman, J., 2009, The MIND program. Available from: https://cdn.flipsnack.com/widget/v2/flipsnackwidget.html?hash=fd1jszu7q&bgcolor=EEEEEE&t=1500654036. [31 July 2019].

[3] MIND catalytic course content. Available from: http://mind.ucsf.edu/mind-course-content. [31 July 2019].

[4] Current participants. Available from: http://mind.ucsf.edu/current-participants. [31 July 2019].

[5] Program description. Available from: http://mind.ucsf.edu/program-description. [31 July 2019].

[6] Impacts of the MIND program. Available from: https://mind.ucsf.edu/impacts-mind-program#hardwork. [31 July 2019].

[7] MIND partners. Available from: http://mind.ucsf.edu/mind-partners. [31 July 2019].

[8] The MIND career exploration road map. Available from: https://mind.ucsf.edu/article/mind-career-exploration-road-map. [31 July 2019].

[9] What is an informational interview?. Available from: https://mind.ucsf.edu/partner-frequently-asked-questions#InformationalInterview. [31 July 2019].

[10] About California Life Sciences Association (CLSA). Available from: https://califesciences.org/about/. [31 July 2019].

[11] Welcome to UCSF connect. Available from: https://ucsfconnect.com/. [31 July 2019].

[12] Transparency in career outcomes. Available from: https://mind.ucsf.edu/impacts-mind-program#Transparency. [31 July 2019].

[13] UCSF Clinical & Translational Science Institute. Available from: https://ctsi.ucsf.edu/. [31 July 2019].

[14] UCSF profiles — find people by research topic or name. Available from: https://profiles.ucsf.edu/search/. [31 July 2019].

[15] Peer mentoring teams for career exploration — design considerations and facilitator training manual, 2018. Available from: https://mind.ucsf.edu/sites/mind.ucsf.edu/files/wysiwyg/MINDFacilitator_training_manual-2018-for-distbn.pdf. [31 July 2019].

myCHOICE: Chicago Options in Career Empowerment

C. Abigail Stayart

Biological Sciences Division, University of Chicago, Chicago, IL, United States

About the program

The goal of the myCHOICE (Chicago Options in Career Empowerment) program is to empower our trainees to select career directions in an informed manner, and provide them with the tools and training to maximize their success in their chosen career. The program covers 10 broad Career Exposure Areas related to the Federation of American Societies for Experimental Biology (FASEB) my Individual Development Plan (myIDP) categories and are presented to trainees through three major types of programming: *Exposure, Education,* and *Experience.* The program is designed to serve trainees throughout the different stages of their academic training and career exploration process and, in doing so, help them to leverage their strong research background.

Our weekly *Exposure* career seminar series, "What can I do with my Ph.D.?" is open to the public and, focuses on a different career type each week. In the span of an hour, the audience is introduced to a panel of individuals with STEM doctoral degrees who have followed a variety of paths to end in roughly the same fields. Approximately 70% of our seminar panelists are alumni sourced from across the country. Over time, we determined that the audience of this seminar series is composed primarily of graduate students in their earlier years of training and postdoctoral scientists in their later years of training. The more than 120 seminars we have hosted since the program's inception have highlighted careers that span the entire spectrum of professional endeavors, including careers in communications (e.g., medical writer, medical science liaison, journal editor, staff writer), outreach and education (e.g., academic librarian, museum curator, high school-level teacher, staff at a non-profit organization), academia (e.g., faculty at research-intensive universities, primarily undergraduate institutions, or community colleges; academic and research administrator; research core facility manager), industry research (e.g., scientist, program manager), business (e.g., biotechnology analyst, finance analyst, general management consultant, entrepreneur), law and policy (e.g.,

BEST: Implementing Career Development Activities for Biomedical Research Trainees
https://doi.org/10.1016/B978-0-12-820759-8.00011-5

regulatory affairs specialist, intellectual patent law specialist, government affairs specialist, science policy advocate, staff at a non-governmental organization), and many, many more.

Our *Education* mini-courses and workshops offer in-depth coverage of a specific subject area or professional development skill. As the program has evolved, the workshop offerings have trended toward a focus on professional development skills (project management, leadership, group management, conflict resolution, negotiations, scientific writing, and improvisation-based communications skills, etc.) because they provide the most universally applicable content that is appropriate for all trainees regardless of their career plans. However, we continue to offer workshops that focus on specific career-related topics, including transitioning to industry research, the drug development pipeline, academic grantsmanship, medical writing, lab and research management, teaching, entrepreneurship, and science policy. We believe that the courses' popularity and low attrition rates can be partially attributed to their timing, as they all take place during evenings and weekends so that trainees' regular research responsibilities are not disrupted. Although we had originally planned that each course would include 12–16 hours of instruction, we immediately learned that our population preferred shorter courses; thus, our current courses range from 4 to 10 hours of instruction. We also discovered that our community's size and low turnover put limits on the frequency with which a course could be offered. Although we do have a few courses that are repeated on a regular annual basis, most of our courses are now offered at ~16-month intervals to avoid lower registration numbers. Finally, we imposed a small registration fee (~\$10) from the beginning to deter people who do not have the serious intention of attending from signing up in the first place.

Experience programming is intended to provide the trainee with a deeper, more intensive experience in a specific career path through a variety of experiential learning opportunities, including part-time unpaid internships, short-term externships, an alumni mentor network, and career-themed "treks," three-day trips to different parts of the country that help expose trainees to different career areas (discussed in detail later in the chapter). These program components were designed for later-stage trainees who are further along the career development pathway and who have identified specific career paths that they would like to explore in more detail. These types of experiences invariably lead to the trainee developing a broader network of professionals working in those fields and to a clearer understanding of the niche opportunities within that career path.

We have achieved particular success through the part-time internships and the treks. More than 75 internships have been completed during their three years of availability; many of these have led directly to offers of employment either through the internship host's company or through connections made during the experience. The part-time format of the internship lends itself well to the trainees' continuing academic research responsibilities and, in fact, interns report that they learned invaluable lessons around time management. Furthermore, our requirement for PI permission enforces the importance of clear communication between trainees and academic PIs about career aspirations.

Similarly to internships, the three-day treks are designed to provide trainees with a nuanced understanding of the broad variety of companies that they could work for. The trek experience helps them to identify specific work environments that may fit their personalities and develop a cultural awareness that they can transmit to potential employers. A significant percentage of trek participants have found job placements in the field they experienced during their trek within a year of their trip by leveraging their new professional networks.

Factors in program success

We credit four major factors for our program's success: (1) it was led by well-known and respected faculty advocates, (2) its components were designed based on identified needs derived from survey data from trainees and faculty, (3) it leveraged existing on-campus resources and collaborators, and (4) its implementation involved the immediate and extensive outreach to and the involvement of graduate and postdoctoral alumni.

Faculty advocates

The involvement of faculty in our program has been essential to the success of both its design and its implementation. Our initial leadership team, made up of three prominent faculty members and two respected and well-connected staff members, had extensive experience in graduate and postdoctoral research training, translational applications of biological knowledge, teaching, and entrepreneurship. Their significant institutional administrative roles provided them the opportunity and authority to influence policy and culture, and their lengthy experience within the latter helped to define realistic goals for the program.

It was critical that key opinion leaders were involved in and continued to vocally support the initiative throughout the start-up process and even at later stages of the implementation of the program. The PI team, consisting of several university officials responsible for the development and oversight of the myCHOICE program, included the Dean of Graduate Students, who immediately began advertising the program during graduate student recruitment and selection. Since 74% of the then-current UChicago graduate student body indicated that an established program such as myCHOICE would have factored favorably in their decision to matriculate here, the program leadership believed that myCHOICE would serve as a strong recruitment tool for prospective students. In fact, three years later, more than half of the matriculating class of 2019 indicated on an entrance survey that existence of the program positively affected their decision to matriculate. To spread the word and excitement for the program, the leadership team obtained support from the Biological Sciences Graduate Program and training grant directors who have continued to support and participate in programmatic elements. Furthermore, it received extremely strong endorsement and support from the divisional dean, the provost, and the president of UChicago.

Program design

When the leadership team first learned of the BEST program announcement in March of 2013, it immediately recognized the program's potential value for UChicago trainees and that UChicago already had a superb infrastructure upon which to build BEST offerings. However, when they set out to prepare the application, they realized how much they did not know about the trainees — their career trajectories, their views about career options, and their views on career preparation — or the faculty, for example. Therefore, in the year preceding the application for the BEST grant in 2014, they assembled a team of UChicago faculty, predoctoral and postdoctoral research trainees, and staff. Working together as colleagues, they first resolved the knowledge deficit through extensive analysis of alumni data and surveying of current trainees and faculty members.

Before embarking on programmatic development, the leadership sought to gain a clear assessment of the perceptions of the current graduate student and postdoctoral trainees regarding career preparedness. A vast majority of respondents (86%) agreed or strongly agreed that "UChicago should make a more concerted effort to prepare students for jobs outside of academia." Our survey also included questions about their career goals, enthusiasm for a myCHOICE-like program, and topics/career paths about which they would like to learn more.

We also queried our trainees on which topics they would be the most interested in learning more about. The most popular topic, listed as the first or second choice by 62% of respondents, was the development of leadership skills. Team building and team dynamics were also highly ranked (listed as the first or second choice by 53%). The high ranking of these topics reflected a desire from our trainees for better professional development, demonstrating how important these skills are regardless of career direction. To address this interest, we proposed a professional and leadership development core curriculum that would intermesh with a career opportunities survey course.

The major career-specific subjects our trainees expressed interest in learning more about were partnering with industry as faculty (29% listed as first choice), teaching either at the collegiate level (24%) or below the collegiate level (9%), business-related topics such as entrepreneurship (21%) and technology commercialization (22%), science communication and writing (17%), science policy (15%), and law-related topics such as intellectual property (9%) and patent law (4%). These interests mirrored the non-academic career paths taken by our alumni.

The role of alumni outcomes-tracking in our program's development cannot be overstated. At the time of the grant's submission, an analysis of the career paths taken by UChicago biological Ph.D. graduates revealed that, on the long term (10 or more years post-graduation), fewer than a quarter of our alumni (∼21%) attain tenure-track faculty positions at research-intensive institutions. Around 55% of our Ph.D. alumni go on to postdoctoral positions (listed as research staff in research-related institutions 0–5 years post-graduation). Many remain as research staff 6–10 years after graduation (∼26%), while others take positions that combine research and teaching (16% for 6–10 years and 29% for 10 or more years post-graduation). Almost half of our alumni, however, are engaged in diverse, non-academic, science-related career paths immediately post-graduation despite the fact that there had been virtually no preparation for or exposure to these "other" career trajectories during their biological research training at UChicago. To best categorize these career trajectories and shape our program, we adopted the 20 myIDP categories, originally devised by FASEB as a framework for postdoctoral IDPs, and used these categories to shape the 10 career exposure areas in our program.

The planning team also understood that a major factor in the success of myCHOICE would be faculty buy-in for the exposure of their trainees to extra-academic career pathways. To assess this sentiment, we surveyed all faculty who were listed as trainers of graduate students or postdocs in biologically-related areas. Of the faculty who responded to our survey (∼40%), there was overwhelming agreement that trainees "whom I mentor should be provided with on-campus opportunities to learn about the full range of career options that use their science training" (94% and 91% for graduate students and postdocs, respectively). Furthermore, 88% of faculty respondents agreed that "new training opportunities to help students and postdocs develop broad skills are acceptable to me if the time they take up is appropriately limited." Importantly, 94% of faculty agreed with the statement that "if my trainees use their science training in a career outside academia I feel I have succeeded as a mentor."

Together, the results of the faculty and trainee surveys showed very strong support for the exposure of students and postdocs to career alternatives outside the academy. An important aspect of our curriculum plan was to have an experiential component whereby trainees could gain hands-on experience through externships and internships in their field of interest. There was considerable faculty support (82%) for short-term (three-day) externships, although support was weaker (59%) for longer-term (one half-day per week for a three- to six-month period) internships. However, 68% of faculty said that their opinion would be more favorable if permission from the PI and/or thesis committee were required for trainee participation. There was less support (51%) for the involvement of trainees in consulting or business ventures (up to 10 hours per week).

Leveraging existing resources

myCHOICE was designed to leverage existing resources at UChicago and to provide a point through which external entities could interact with trainees within the Biological Sciences Division (BSD). Up until this point, many existing resources, such as UChicago's world-class professional schools (for example, the University of Chicago Law School, the Harris School of Public Policy, and the Booth School of Business, which is regularly ranked #1) had been unknown or vastly underutilized by our biological sciences trainees.

Specific elements of our programming were therefore designed to create connections with these bodies and to take advantage of their wealth of expertise. For example, the Writing Program, which provides writing assistance to all UChicago divisions, has a flagship course called Little Red Schoolhouse (LRS), which focuses on helping advanced writers structure complex data, develop extended arguments, and position their work as a contribution to ongoing debate in their fields. The course is offered in several different versions for graduate students, M.B.A students, professional students, and advanced undergraduates; however, the full-quarter course requires a significant time commitment from participants and is entirely unavailable to postdoctoral scientists. We worked with the directors of LRS to develop a shorter version of their course that required a smaller time commitment and was also open to postdocs.

Similarly, the Booth School Leadership Excellence and Development (LEAD) program focuses on enhancing self-awareness around critical leadership skills, and encourages students to take a personal and active role in the ongoing development of these skills. The full-length course is only open to students of the Booth School, so a shorter version was created to tailor content to myCHOICE trainees in the biological sciences.

A third example is provided by the Chicago Center for Teaching (CCT), which has long been available to all graduate students and postdocs. The CCT has a teaching certificate available that can be obtained by completing a series of workshops and a subject-specific pedagogy course, and allows participants to receive feedback on their teaching techniques in real-world situations. However, the CCT was rarely used by trainees from the biological sciences, many of whom did not know it existed, and was generally perceived to be "not for scientists." The myCHOICE team worked with CCT to clearly tailor their extensive workshop offerings to our trainees, created new myCHOICE-specific mini-courses, and helped to spread the word about the opportunities provided by the center.

Beyond formal programs and offices, our invaluable collaborators include the central University Graduate Affairs Office, the tech transfer office, the very active Postdoctoral Association, the Graduate School's Dean's Council, and burgeoning student-run groups. Since the inception of myCHOICE, the list of these initiatives, whose interests align with the myCHOICE mission, has continuously grown and now includes a biotech association, a consulting club, a local Society for Advancement of Chicanos/Hispanics and Native Americans in Science (SACNAS) chapter, the Graduate Recruitment Initiative Team (GRIT), and more. These partners not only provide resources and insightful suggestions for new program components, but they also have become a word-of-mouth network that helps to spread awareness of myCHOICE to all corners of the BSD and other STEM groups on campus and in the broader Chicagoland area.

Alumni involvement

The success of myCHOICE is derived not only from its curriculum designed to expose and educate our trainees to diverse career pathways, but from the establishment of a *vital network of external and internal mentors* that partner with our trainees to provide them with real-world experiences. Our mentors come from diverse backgrounds and careers and are UChicago alumni, business partners, or others who are interested in helping trainees transition into their career of choice. Many of our mentors have enthusiastically agreed to participate in our curriculum development, drawing on their vast expertise to educate our trainees. The myCHOICE program provides a robust infrastructure, but the mentors are the actual connective tissue of our career development efforts as they provide our trainees in the BSD with important links to their areas of expertise.

Alumni involvement in all stages of our programming is a significant component of our program's success. We identify alumni through a variety of sources, including our Alumni Relations & Development Office, departmental administrators, personal connections, LinkedIn networking, and Internet search engines. In most cases, outreach to an alumnus starts with a simple email that introduces the program and invites them to visit campus as a panelist or alumni mentor. For some alumni, this is the first time that they have been contacted by a university office since leaving training and they are both surprised and flattered to learn that their unique career trajectory is the reason for the contact.

Invariably, alumni report that they "would have loved to have a program like this when I was here," and are excited to provide a form of mentorship to current trainees that they, themselves, did not enjoy during their training years. Current trainees can benefit a lot from engaging with the alumni network. They have reported, for example, increased confidence in their ability to transition away from academia, a greater sense of connection with people who have "gone through the same institution," and informational interviews or even job offers stemming from their alumni connections.

We encourage other institutions to consider the immense value of regarding former post-doctoral trainees as alumni. Although the traditional definition of an alumnus is often restricted to those who have been registered in a course at an institution, we broadened our definition to include postdoctoral scientists. This new definition reflects the high value of these later-stage trainees who contribute to our community for a length of time that is often equivalent to that of a doctoral student. By offering a career development program that specifically considers their needs, an institution imbues in its population of young professionals a sense of proud affiliation with the institution that offered them support during a critical time in their professional development. UChicago postdoctoral alumni are among our staunchest supporters because, at the moment they understood that their research training had not entirely prepared them to make their next step, myCHOICE was there to supplement their experiences.

Notable pieces of programming

We would like to highlight four pieces of programming that have been particularly popular and have a demonstrably high impact on our campus.

"What can I do with my Ph.D.?" career seminar series

Many BEST institutions developed a seminar series intended to expose their community to the wide variety of career trajectories that value an advanced degree in a STEM field. Through our post-event surveys, we learned fairly quickly that, at seminars with only one speaker, we risked losing audience approval based on the quality of the speaker and the perceived relevance of that person's career trajectory to the individual trainees; audience members sometimes tuned out if they felt that the speaker did not reflect what they believed about that career path. Therefore, in the second year of our program, we shifted the format to small panels of speakers from similar careers but who individually took unique paths to get there. The panel format allowed us to underline the fact that there are multiple paths leading to the same end, explore the nuances of similar professional roles, increase the likelihood that an audience member would personally identify with at least one of the panelists, expand the potential network for interested trainees, and engage more alumni. Although recruiting and coordinating a diverse panel requires more attention and time than required for identifying a single speaker, the value added for the trainees ensures that they leave the event feeling that it was a worthwhile use of their time.

Beyond the bench: the business of running a lab

One of our most popular mini-courses among our academic-track trainees and postdocs highlights aspects of running an academic research lab. We cover topics that are often not explicitly discussed between mentor/mentee and that are often left out of formal training. Four 90-minute panel discussions, moderated by myCHOICE staff, focus on the following topics: The Application Process, Negotiating, and Budgeting Your Start-up Package;

Acclimation as a New PI; Getting Your First Grant; and Hiring, Firing, and Managing Your Team. Eight faculty members representing diverse fields and career stages are recruited to participate in a minimum of two sessions in order to provide consistency across the mini-course. The moderator focuses the first 45 minutes of each session around specific questions that all panelists answer, which is then followed by 45 minutes of Q&A. Ideally, the faculty members would meet as a group prior to the start of the course to discuss the topics and critical content they would like to address; however, in the absence of that group meeting, it is important to offer each session's panelists a list of questions in advance to help them prepare for the discussion.

Career-themed treks

A "trek" is a three-day trip to a location where there is a high concentration of jobs in a specific career field. For example, the Biopharma Trek takes trainees to Boston, the Computational Biology Trek to the San Francisco Bay Area, the Science Policy Trek to Washington, DC, and the Financial Services Trek to New York City. During each day of the trek, we make three to four site visits at companies that are chosen to represent the breadth of career opportunities in a given landscape. We make a deliberate effort to include companies of different sizes (e.g., ranging from 2 to 2,000 employees), life spans (e.g., start-ups or decades-old market players), and cultures (e.g., a 9-to-5 work day or a work-from-home expectation) so that trainees come away with a more nuanced understanding of the featured industry and of where they might most comfortably fit within it.

Participants are selected through a competitive application process; eligibility is restricted to graduate students in their final years of training and postdoctoral scientists in their second year of training or beyond. The application process is designed to select trainees who are clearly at the stage of serious contemplation of the career path, and requires three mock cover letters for jobs at the companies that will be visited on the trek, a resume, and a personal statement that discusses their interest and past evidence of their participation in activities related to the trek focus. We recommend that potential trek applicants use the University's Office of Career Advancement to receive feedback on their application documents prior to submitting them, which ensures not only that our trainees become aware that the service exists, but that they engage in a professional development activity through the application process even if they are not selected for the trek. The application sets a high bar for participation, which dissuades potential applicants who are not seriously considering that career path and results in a cohort that is highly prepared for the site visits.

Although we had anticipated that the trek format would carry unique benefits for trainees by providing intensive exposure to a potential career path, perhaps the most surprising element of the treks' success has been the opportunity to engage with alumni across the country. During the planning process, we use our extensive alumni database as well as publicly sourced information to identify alumni who are working in the selected location and reach out to them to gauge their interest in participating in the trek. Some alumni are excited to coordinate site visits, while others appreciate the invitation to attend an alumni reception. Our treks have served to re-establish the visibility of the university with individuals who may have left decades ago and who were discouraged by the actual or perceived lack of support from their PI to follow a path that led them away from academia. Engaging with myCHOICE and our carefully selected trainees improves their general sentiment toward the University of Chicago and has the potential of increasing philanthropic relationships.

In summary, although the cost per person is higher than any other activity in our program (myCHOICE covers trainees' travel and lodging expenses), the immediate positive effect that it has in the career development of our trainees and the relationships built across the country make myCHOICE treks an invaluable investment.

Part-time internships

A key goal of myCHOICE *Experience* programming is to provide real-world, practical experience in specific career fields. Internships, hands-on opportunities of limited duration (normally weeks to months) are an important component of this training experience. myCHOICE teamed with on- and off-campus partners to provide a diverse array of internships in environments that value and leverage trainees' scientific training. Two distinguishing features of our internships are their length (10 weeks) and their part-time structure (10 hours per week), which accommodates the full-time research responsibilities of graduate students or postdoctoral scientists. All of our internships are unpaid, thereby allowing trainees of all nationalities and types of appointments to take advantage of the opportunity for experiential learning outside the lab.

The development of the nature, structure, and content of our internship program was based on survey feedback from trainees and faculty. We learned that trainees recognized the worth of exposure to the "outside world" and desired opportunities to explore a wide variety of careers, including industry research, entrepreneurship, healthcare, communications, regulatory affairs, teaching, research administration, and technology and commercialization. However, trainees were hesitant to pursue traditional full-time internships due to the imposition it would place on their research and, for graduate students, to the potential delay in the completion of their degree. Some trainees also expressed concern that doing an internship would signal to their PIs or to potential employers that they were no longer interested in an academic research career when, in fact, many of them simply saw the value in broadening their understanding of the workforce outside of academia and in diversifying their personal skillset.

We learned that faculty also recognized the potential value of an internship experience for trainees who would need those connections in order to transition away from academic research. Faculty members were, however, hesitant to give unconditional approval to internships, stipulating that their support was contingent upon two factors: (1) an appropriate limitation to the time commitment, (2) the involvement of the advisor in granting permission for the trainee's participation in an internship. Out of respect for these concerns, we developed an internship program that is carefully curated, narrowly focused, and that includes a formal Internship Learning Agreement (ILA) that clearly recognizes and manages the expectations of all the relevant stakeholders (the intern, the internship host, and the intern's academic mentor).

One of the biggest challenges and most time-consuming aspects of building our internship program was identifying opportunities that would yield a valuable experience to both intern and internship host. To that end, all potential hosts are extensively vetted and assisted in developing an internship project that will benefit their office or organization and that will add value to the intern. It is critical that the host understand that the internship is intended to be co-curricular and that the trainee's research and training progress must remain a top priority; therefore, projects must be selected based on the appropriateness of the time commitment required from the intern. It is also important than an internship project be designed around a specific deliverable; the product becomes something that a trainee can reference on their resume as evidence of the experience they gained and that the internship host can appreciate as something that might not have otherwise been generated. The deliverable is advertised as part of the internship description, and each host/intern pair develops the project and documents it in more detail in the ILA prior to the start of the internship.

In addition to a requiring a detailed description of the project, the ILA accomplishes several other functions. As we believed that it was important that the trainee be formally granted the time to commit to a project outside of their research, the ILA requires documentation of acknowledgment, approval, and support of the trainee's time commitment to the internship from both their academic advisor and the Dean for Graduate Affairs. We recognize that the stipulation of advisor approval may create an insurmountable barrier for some trainees who may not have the benefit of an open dialogue about their career trajectory with their advisor. However, we also understand that the success of an internship program hinges on the support of all stakeholders in our community and, thus, feel strongly about the inclusion of this requirement. The process of obtaining permission ensures that the trainee and mentor engage in a conversation about the trainee's professional development, therein building a stronger and more communicative trainee/trainer relationship.

Our internship placement process involves a competitive application; applicants are not guaranteed an internship, nor are internship hosts obligated to move forward with an applicant simply because they applied. We believe that the application process itself is a valuable experience for our trainees regardless of whether or not they are ultimately selected for the internship. The application plays two important functions: 1) it increases the likelihood that the applicant will avail themselves of the Office of Career Advancement to receive feedback on their application materials or to conduct a mock interview, thereby encouraging the type of preparation than an applicant should due to prepare for the job market but in a low-stakes environment, and 2) it permits the host to truly vet their applicants to ensure that they are selecting someone who is dedicated to the relationship even in the absence of monetary compensation.

Arguably, the most challenging part of growing an internship program is identifying the right internship hosts. A great place to start is to reach out to campus partners who would clearly benefit from the scientific knowledge that an intern would bring. Examples include communications and media relations offices, tech transfer offices, product accelerators and incubators, core facilities, administrative offices, outreach programs, and alumni development offices. At this time, all of our off-campus internships are facilitated by alumni or individuals who have already engaged with myCHOICE as seminar panelists or site visit hosts. These sites have included museums, science non-profits and advocacy organizations, educational organizations, public sector offices, tech start-ups, consulting groups, and capital investment firms.

Lessons learned

Through analysis of general community surveys and event attendance, we can definitively conclude that myCHOICE is positively affecting the training experience at the University of Chicago and that, broadly, trainees feel more comfortable initiating conversations about their career trajectories with their peers and mentors. If the overarching goal of this type of career development support is to increase our trainees' confidence about their futures and help them to launch into their careers of choice, then we have indeed succeeded.

However, in spite of the careful tailoring of the program to our local culture, the success of the program was not guaranteed and strategic alterations were required to adjust to unexpected factors. For example, while we had originally planned to use off-campus locations for events, we quickly found that trainees did not attend these events, probably due to the additional travel time and the inconvenience of finding transportation. Another example relates to potential on- and off-campus collaborators with whom our connection was a single individual, which made those relationships susceptible to termination if that individual moved on to another job. We also underestimated the effort required to implement and maintain such a broad program; within one year of program inception it was necessary to involve a second employee in the daily execution and publicity campaigns associated with our many events.

Several lessons have become apparent now that multiple years of data are available. myCHOICE has closely monitored and analyzed event attendance records, not only for the number of attendees, but also for attendees' demographic information, stage in training, departmental affiliation, and the number and type of events that each one of them attended. These data reveal a shift in the number of graduate students who participate, from 38% during the first full year of programming to 25% in the program's fourth year. In contrast, we have seen steady 20% participation rate from the postdoctoral community. Closer analyses of the data reveal what we believe to be a "pent-up demand" phenomenon in which myCHOICE was novel to every community member in our first year but which created a population of older graduate students who no longer needed the same content — namely, a basic exposure to the diverse career paths taken by Ph.D.-trained scientists — by their fourth year. We are beginning to observe a normalization of attendee numbers according to their stage of training as was originally intended by the program's designers; early-stage trainees or those who are new to campus are the primary attendees of the seminar series, but later-stage trainees instead participate in activities that reflect a pre-existing familiarity with their career options, such as internships, specialized mini-courses, and events that involve networking with alumni.

Also related to attendance data was the recognition that there is decreased demand for specific types of mini-courses when they were offered in sequential years. We have interpreted this decrease as a consequence of our community's turnover rate; when an event is offered more than once per year, there simply are not enough new trainees on campus to reach reasonable attendance levels. As stated above, we have found that 16-month intervals seem to be more suitable for our size and turnover rate.

Finally, we have recently started to strategically address the community's misunderstanding that myCHOICE is "only" an "alternative careers program" by developing more broadly-applicable mini-courses, with subjects such as negotiation, team management, and individual leadership development. We are now emphasizing the broad relevance of this programming to the faculty community, with the intention of demonstrating that myCHOICE is, in fact, a program for all trainees at the University of Chicago regardless of their intended career path.

Final thoughts

NIH funding allowed us to realize a robust and meaningful career empowerment program for biological sciences graduate students and postdoctoral trainees at UChicago, and it revealed a considerable need and appetite for these offerings throughout the STEM community on our campus and at other academic institutions across Chicagoland (such as Northwestern University, Rush University, and the University of Illinois at Chicago), whose trainees have also participated in our programming. Indeed, several of UChicago's divisions have expressed interest in the wholesale adoption of our structure to meet the needs of engineers and physical and social scientists. The outgrowth of these efforts has been a series of strategic conversations about pan-university and pan-Chicago partnerships. We envision a future of continuous growth and improvement of our original model, extended well beyond the scope and audience conceived by our original proposal to the NIH Common Fund that supported the Broadening Experiences in Scientific Training mechanism.

Acknowledgments

The myCHOICE primary investigators currently include Erin Adams, Professor of Biochemistry and Molecular Biology; Ellen Cohen, Executive Director of the Center for Health and the Social Sciences; Victoria Prince, Professor of Organismal Biology and Anatomy; and Julian Solway, Professor of Medicine and Pediatrics and Director and Associate Dean of the Institute for Translational Medicine. Alan Thomas, during his tenure as the Associate Vice President and Director at the UChicago Center for Technology Development & Ventures, was a founding primary investigator and played a critical leadership role in the conceptualization and launch of the program; he continues to serve on the program's external advisory board. The program's administrative team has included former Project Coordinator Marissa Pelot, Communications Manager Christopher Peña, and Program Director Abigail Stayart. The myCHOICE team greatly appreciates the support of our institutional and Chicagoland partners. UChicago GRAD staff members Michael Tessel, Senior Director of GRAD Career Development and Employer Relations, and Briana Konnick, Associate Director of Graduate Career Development, make essential and generous contributions to our program. We are deeply grateful to our vibrant alumni community and vocal supporters across the country. myCHOICE was funded through NIH grant number 5DP7OD020316-05.

Better through BESST!

Inge Wefes

University of Colorado Denver, Anschutz Medical Campus, Aurora, CO, United States

Introduction

Over the last 20 years, it has become increasingly clear that the number of trainees in the biomedical sciences far exceeds the number of tenure-track positions that are needed to employ all of them in academic research positions. Some universities had already responded to this realization in the late nineties by inviting experts in various non-academic but science-related jobs to provide pre- and postdoctoral fellows an overview of the tasks and challenges related to their professions. Others, like the University of South Florida (USF), initiated extended training opportunities that allow doctoral students in medical sciences to concurrently pursue a Ph.D. plus a master's degree or graduate certificate in disciplines such as Business Administration, Public Health, Biomedical Engineering, Bioethics, Bioinformatics,

BEST: Implementing Career Development Activities for Biomedical Research Trainees
https://doi.org/10.1016/B978-0-12-820759-8.00012-7

and more. This Ph.D. PLUS Program (developed in 2000 under Inge Wefes's leadership) has the ultimate goal of increasing the competitiveness of Ph.D. trainees for careers inside and outside academia. Over time, funding agencies started to support a broader biomedical training, and in 2013, the National Institutes of Health (NIH) launched the first 10 grant awards for **B**roadening **E**xperiences in **S**cientific **T**raining (BEST). The Graduate School at the University of Colorado Denver|Anschutz Medical Campus was one of the first NIH BEST Award recipients (PI: Wefes) and shares here some of the highlights of its successful program.

Where we started

The University of Colorado Denver (CU Denver) has two campuses, the Denver Campus and the Anschutz Medical Campus; the Graduate School is located on both. The NIH BEST Program is hosted on the Anschutz Medical Campus and serves about 350 doctoral students in the biomedical sciences and about an equal number of postdoctoral fellows.

In 2013, when the BEST Program started, the Graduate School was mainly involved in administrative tasks, such as admissions and graduation, and the Postdoctoral Office was mostly concerned with issues related to human resources. A Postdoctoral Association existed, but it did not have sufficient support. Furthermore, there was a student organization called Alternative Careers in Science Club that invited speakers to address career topics beyond the traditional faculty career. When CU Denver earned the BEST Award, the program director, Wefes, began collaborating with these organizations, and the BEST Program provided some much needed and appreciated support and resources.

What we did

Creating awareness

Through presentations to students, postdocs, faculty, and department chairs, the CU BEST Program enhanced the institutional awareness that student and postdoc training had to be broadened beyond just preparation for academic research in order to make trainees more competitive for science jobs both inside and outside academia. To benefit the largest number of students, the CU BEST Program opted against a cohort model and advertised all career development opportunities to the biomedical sciences Ph.D. students and postdocs on the Anschutz Medical Campus.

Framing the program

CU Denver's BEST Program is structured along the life cycle of a Ph.D. student in the biomedical sciences. As such, it starts each academic year with the annual **Graduate and Professional Skills (GPS) Orientation** for incoming Ph.D. students and ends each academic year with the annual **Milestones of Success Celebration** for all trainees on the Anschutz Medical Campus. The GPS Orientation spans two days and is divided into two stages. First, it provides a half-day session on basic information about course registration, student insurance, biosafety, the Student Senate, etc. Second, it divides students into three groups to engage in mini-workshops on topics such as career options, the value of the Individual Development Plan (IDP), self-assessment with the Gallup StrengthsFinder, writing an NSF Graduate Research Fellowship application, mentor/mentee relationships, emotional and mental health, critical and analytical thinking, and more. Each of these interactive workshops lasts about 45 minutes and is instructed by faculty or staff members from the Denver and Anschutz campuses.

The Milestones of Success Celebration is held at the end of the academic year and celebrates various milestones of the students' and postdocs' progress, i.e., who has graduated, found a job, published a paper, finished qualifying exams, earned a grant or fellowship, and more. At this time, we award the **International Visiting Scientist** fellowship to one Ph.D. student and one postdoc. With this award, trainees can spend four to six weeks in a laboratory outside the USA to learn new techniques, use equipment that might not yet be available in their home labs, build and foster relationships, and become familiar with the international enterprise of biomedical sciences and biotechnology.

During the Milestones of Success Celebration, we also issue **Trainee Peer Mentor** and **Faculty Sponsor Awards**, which recognize the professional and career development support that peers or faculty provide. Thus, in addition to honoring the awardees with this recognition, we use this celebratory occasion to instill in attendees the spirit that professional and career development training is an important part of the training in the biomedical sciences and that faculty who support these efforts deserve to be especially recognized.

In early 2016, it became clear that in many cases the term "science" in "biomedical sciences" is often too narrowly associated with bench research. To be more inclusive and better reflect the many different processes and procedures of the biomedical sciences, including computation and data sciences, we renamed the BEST Program at CU **BESST Program**, which stands for **B**roadening **E**xperiences in **S**cientific and *S*cholarly Training. Needless to say, this name would allow us to later apply the successful program, with its career development opportunities, to a greater student body in the broader liberal arts and sciences.

Creating career development opportunities

Skill-building workshops

Between the two bookends — the GPS Orientation and the Milestones of Success Celebration — CU's BESST Program offers interactive workshops designed according to

FIG. 12.1   BESST Program career development opportunities.

skills that employers often report to be deficient in trainees who join the workforce fresh out of university. These professional and career development workshops are divided into two main subject areas: (1) Thinking & Communicating and (2) Leading & Managing (Fig. 12.1).

Area 1 includes workshops such as **Critical & Analytical Thinking, Scientific and Technical Writing,** and **Principles & Strategies of Effective Teaching**. Area 2 includes workshops such as **Science in Industry, Life Science Development & Commercialization, Team Building & Leadership Development** and **Project Management**. Although each workshop has its own focus that justifies its inclusion in one of the two subject areas, workshops often indirectly also help train other skills, as indicated by the connecting lines in Fig. 12.1. For example, the Life Science Development & Commercialization workshop or the Project Management workshop also have to address team building and leadership skills, even though we found it valuable to also offer a separate workshop with a focus on just those skills. Similarly, the Principles & Strategies of Effective Teaching workshop also addresses issues related to speaking and presenting.

With the exception of the workshops Science in Industry and Project Management, all workshops cover a total of 15 hours (6 sessions of 2.5 hours each), and are usually held over lunchtime to be the least competitive with the trainees' research time. Not only do the 15 hours provide the necessary time for practicing the newly acquired skills, but this timing was also selected with the perspective of future sustainability, as 15 hours corresponds to the contact hour requirement of a 1-credit graduate course for which the instructors could be paid out of the tuition revenue.

The Science in Industry workshop spanned two full days and covered practical issues such as deciding on a CV versus a resume, informational and job interviews, salary negotiations, setting up a LinkedIn account, and more. The Project Management workshop lasted four full days, distributed over four weeks. With this workshop, participants obtain the minimum required training to apply for testing at the Project Management Institute to become a Certified Associate of Project Management (CAPM), a certification that is internationally recognized and is valid for five years.

Workshop instructors were mainly faculty recruited from both campuses. For example, for the Critical & Analytical Thinking workshop, we recruited a faculty member from the Philosophy Department. For the Writing workshop, we invited the support from a faculty member of the English Department who had formerly also worked as a science and grant writer in a company. The teaching workshop was offered by a team of faculty members who had not only extensive teaching experiences on their own, but also substantial knowledge in educational theory and pedagogy. The Life Science Development & Commercialization workshop was developed and delivered by a member of the Business School who also had a great number of connections to local biotech companies from which he recruited experts for selected topics for the workshop. The Team Building & Leadership workshop was offered by a senior psychologist with extensive experience in corporate coaching.

All workshops and all workshop sessions started with a five-question self-assessment of the skills and knowledge that were to be addressed in that specific session (pre-event assessment), and all workshops and all workshop sessions ended with the same five questions (post-event assessment) in order to evaluate the participants' self-efficacy (their confidence that the training prepared them to be better in the tasks that were taught in that workshop session); it has been known for a long time that increased self-efficacy leads to better performance. To keep the workshops fun and relatively stress free and to avoid the attrition of people who are never short of work in their laboratories, we did not use any exams or other more objective assessment measures to evaluate the effectiveness of the workshops. Fig. 12.2 illustrates pooled answers for just two questions for three of the workshops we offered.

Testimonials from trainees, especially those in which they attributed their successful job search to one of our workshops, have also reinforced the value of these activities. An example of one such testimony, in this case related to the Principles & Strategies of Effective Teaching workshop, informally called "Learning How to Teach" (LHTT), follows:

> I took the LHTT Workshop in 2016 as a graduate student. Although I had previous teaching experience, this workshop was my first exposure to active learning techniques. During my interview for the lecturer position, they specifically asked for examples of active learning techniques that would be beneficial for the courses that I would be teaching. If I hadn't taken the LHTT workshop, I wouldn't have even known where to start. Plus, one interviewer asked me to verbally communicate my teaching philosophy, and I would not have been able to do that if I hadn't drafted a teaching philosophy in the LHTT workshop. The hiring committee told me (after the fact) that although I was the most junior applicant, it was evident that I had put more thought into my teaching than any of the others. I know without a doubt that the knowledge and experience I gained from LHTT are the reason that they felt that way.

We use short, follow-up surveys six months to a year after the training to inquire with trainees if they think the training had an impact on their career preparedness. An example of four questions in such a follow-up survey for the Speaking & Presenting workshop is illustrated in Fig. 12.3.

Workshop: Advanced Critical and Analytical Thinking

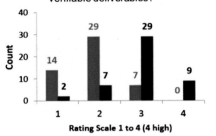

How knowledgeable are you about how to create verifiable deliverables?

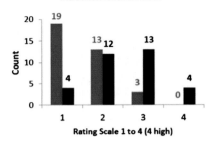

How confident are you that you can understand why induction is fallacious?

Workshop: Scientific Writing

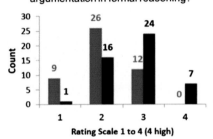

How confident are you that you can *explain* the difference between valid (good) and invalid (bad) argumentation in formal reasoning?

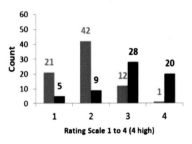

I can use basic narrative principles to draft, asses, and revise the sections of my technical reports.

Workshop: Project Management

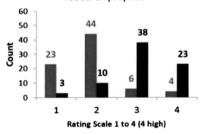

I can effectively asses my writing to determine if my research proposal meets the guidelines for a strong research proposal.

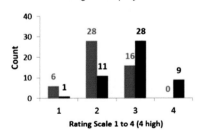

How confident are you that you could identify when and how it would be possible to make productive changes to a project?

FIG. 12.2   Selected assessments of increased trainee self-efficacy. Red, pre-assessment. Black, post-assessment.

Networking and community building

In addition to skill-building workshops, the CU BESST Program invited trainees to learn about a variety of topics related to professional and career development through seminars such as Intellectual Property, Personal Strength and Career Selection in the Biomedical Sciences, and From Biochemist to Consultant, and panel discussions with members of local

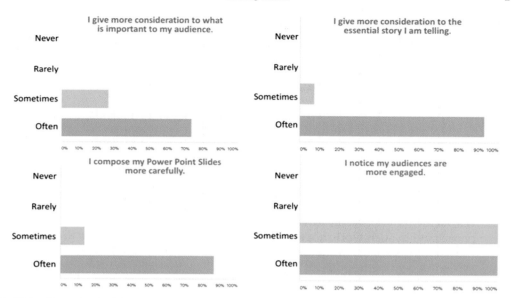

FIG. 12.3 Example of a follow-up survey to assess the long-term impact of the Speaking & Presenting workshop.

industries. To respond to the increasing demand for more data science training, in 2014, CU BESST set up an **Annual Spring Mini-Symposium** called **The Power of Informatics to Advance Health** for trainees and faculty. In addition to the exchange of new basic and clinical data in bioinformatics, these symposia serve as networking events and facilitators for new collaborations.

Building on the momentum of this symposium, the BESST Program initiated and supported the development of a **Graduate Certificate Program in Biomedical Data Science** that started its first enrollment in Fall 2017. Biomedical data science has become a central component of biomedical research, and biomedical researchers with data science knowledge are advantaged on multiple fronts: they are familiar with computer programming; they can locate, access, analyze and visualize biomedical data sets; they can understand and apply various machine learning techniques and data analytics to solve real world biological problems; and they can communicate effectively with other biomedical researchers and computational data analysts in a team science environment. The Graduate Certificate Program in Biomedical Data Science has attracted graduate students, postdocs, and faculty.

Training trainers

To this day, many faculty members themselves have never received any training in teaching, writing, speaking, project management, etc., as their own training fully focused on the development of their subject expertise in a special arena of biomedical sciences. Over time — in some cases, encouraged by their own trainees — some faculty members attended CU BESST workshops, even though they were not the primary target audience. For example, one faculty member who participated in the Scientific Writing workshop sent a note to the BESST Program Director stating:

> For me, the BESST writing workshop was a fantastic opportunity for hands-on writing practice and feedback. The instructor broke down the writing elements to workable parts and changed my whole approach to scientific writing. I highly recommend this workshop for anyone who needs to obtain grants and publish in scientific journals.

Other faculty members have also participated in the Critical & Analytical Thinking workshop, indicating that some form of training in this area can be valuable even for faculty. It soon became clear that providing some more concentrated and shorter training for faculty (compared to what is provided to pre- and postdoctoral trainees) could serve three purposes: (1) faculty might overall become better mentors, (2) they might be better trained to practice these professional skills with their trainees, and (3) after having obtained this training now themselves, faculty could become advocates who not only encourage their students and post-docs to participate in such workshops, but who would also be potential partners in BESST's sustainability efforts to convert some workshops into credit-bearing courses and make them an integral component of the curricula of the various doctoral programs on the CU Anschutz Medical Campus.

To this end, we offered, for example, a condensed workshop on speaking and presenting to seven senior faculty members who had all given countless presentations before. Not only did all of them find the workshop helpful for themselves, but, in the workshop evaluations, they also all shared that they thought that the workshop would indeed be helpful for doctoral students and postdocs as well as junior and even other senior faculty! We also offered a condensed version of the Project Management workshop to 10 junior faculty, and not only did they all find value in this training for themselves, but they also agreed that it would be valuable or very valuable for Ph.D. students and postdocs and even junior and senior faculty. Indeed, those faculty members who used the opportunity to participate in one of the short, faculty-only BESST Program workshops have become strong advocates for the BESST Program and related training opportunities.

What we achieved

Based on discussions with faculty and trainees and follow-up surveys, the CU BESST Program has increased awareness about the usefulness of and the need for career development opportunities for pre- and postdoctoral fellows in the biomedical sciences. BESST helped to abolish on the Anschutz Medical Campus the term "alternative careers," as this term not only has the connotation of being a slightly inferior "Plan B," but based on the numbers of alumni employed in academia, it is indeed the tenure-track appointment that has become the exception or alternative. The involvement of faculty in opportunities for training as well as instructing — for example, in the GPS Orientation workshop on mentor/mentee relationships — has made faculty members active supporters of BESST activities. These attitude and culture changes have been further enhanced by awards that recognize faculty as sponsors, i.e., that they do not just serve as academic mentors and subject experts in the biomedical sciences, but are also engaged supporters of their mentees' career plans.

The BESST Program has created many valuable and successful career development opportunities, including workshops, symposia, networking opportunities, panel discussions, and a graduate certificate program in biomedical data science, that make it difficult for trainees to avoid the awareness that career development has to start early and should not be left to the time when a job is needed. With these constant opportunities and stimulations, BESST also enhances in trainees the sense of self-responsibility: even though they are not yet independent researchers, they can already take components of their career development into their own hands. In this spirit, and as a BESST Program spin-off, pre- and postdoctoral trainees on the CU Anschutz Medical Campus initiated in 2014 the **Academic Industry Alliance (AIA),** which strengthens connections between academia and local industries. They now organize on the Anschutz Campus an **annual industry showcase** where companies occupy booths and present on their products or services. In addition, the AIA also organizes the annual **Rocky Mountain Biotechnology Symposium,** which includes short presentations from local industry representatives and networking opportunities that have led to numerous cases of employment of students and postdocs.

Another spin-off of the BESST Program is the *Ph.D. Post,* a fall- and spring-semester newsletter by and for graduate students and postdocs to inform each other and the Anschutz community about new developments and opportunities inside and outside academia, share experiences and career advice, and voice opinions about new campus features or educational policies. As an additional means of honing their communication skills, trainees also set up a **Speaking & Presenting Club,** where they keep practicing — with peer critique and support — the mechanics of effective speaking.

Where we will go from here

While the NIH BEST funding is not renewable, the BESST spirit will remain! Many organizations on CU's Anschutz Medical Campus have enhanced career and professional development opportunities for trainees, and the generous funding by the NIH for the BESST Program allowed for valuable experimenting and fertilizing of these efforts.

So far, four of the CU BESST workshops — Scientific Writing, Principles & Strategies of Effective Teaching (LHTT), Project Management, and Speaking & Presenting — have been converted into course-shops, i.e., credit-bearing courses in which people who are not interested in transcripted credits (such as postdocs or junior faculty) can enroll, for the same price and service, as if the course were a workshop. The Leadership & Team Building workshop will be continued by the Colorado Clinical and Translational Sciences Institute, a major NIH-sponsored translational sciences center that also has an educational mission. The Center for Personalized Medicine, an Anschutz biomedical data science center that analyzes genetic and genomic data to find personalized disease preventions and interventions, will now host the annual conference The Power of Informatics to Advance Health. Ph.D. programs have started to allow their students to concurrently enroll in the new Graduate Certificate Program in Biomedical Data Science (reminiscent of USF's Ph.D. PLUS program), thereby giving their students a competitive advantage for jobs inside and outside academia. The training on life science development and commercialization is continued in a graduate course called Building Biotechnology. In addition, the CU Denver|Anschutz Office of

Innovation now offers more short training modules on innovation and commercialization. Efforts are also ongoing to find new funding to allow trainees to spend some time abroad as international visiting scientists.

While the BESST Program and its spin offs, such as the AIA and the *Ph.D. Post*, have clearly made a lasting impact on the Anschutz Campus of the University of Colorado, the spirit and the efforts of CU's BESST can also be applied to disciplines outside the biomedical sciences, and the Graduate School at CU Denver|Anschutz Medical Campus will continue to make career development — university wide — an integral feature of graduate training programs. CU is better through BESST — but the BESST is not enough!

Acknowledgments

The BESST Program is most grateful for the generous support by the NIH, grant number 11393977. In addition to many unnamed faculty and trainees as well as presenters of seminars and contributors to panel discussions and networking opportunities, CU's BESST Program owes special thanks to Judith Albino, Mark Bauer, Matthew Berta, Tiffany Dahlberg, Michael Ferrara, Jim Finster, Patricia Goggans, Mitch Handelsman, Laurel Hartley, Chris Hill, Kenneth Jones, Brian Meara, Arlen Meyers, Christopher Miller, Susan Nagel, Larry Petcovic, Tzu Phang, Christopher Phiel, Randy Ribaudo, Alle Rutebemberwa, Aik-Choon Tan, Rose Shaw, and Emily Wortman-Wunder.

13

Creating the "new normal": Career development embedded into the Ph.D. curriculum for all trainees

Cynthia N. Fuhrmann, Spencer L. Fenn, Daniel Hidalgo, Brent B. Horowitz, Heather S. Loring, Sumeet Nayak, Meghan E. Spears, Grant C. Weaver, Mary Ellen Lane

Graduate School of Biomedical Sciences, University of Massachusetts Medical School, Worcester, MA, United States

How can we, in the short span of the NIH BEST award, seed a paradigm shift that will prompt and sustain students' proactive career planning early in Ph.D. training, and do so in synergy with thesis research? In the Graduate School of Biomedical Sciences at University of Massachusetts Medical School (UMassMed), we decided to take an approach fundamentally different from traditional career development programs: rather than focus on adding opt-in workshops or opportunities, we integrated career and professional development directly into and across the span of Ph.D. training as part of the required curriculum for all students. Five years later, we are now seeing significant changes on our campus, from students embracing career planning and taking earlier actions toward their career development, to changes in our institutional culture. Here, we tell the story through student voices.

Why we took a curricular approach

Prior to receiving a BEST award, our campus had few central programs for career and professional development. As a medical school, we did not have the types of services typically available on comprehensive university campuses, such as a career center, writing center, business school, or center for teaching and learning. Several years ago, recognizing that we needed to more explicitly address the professional development needs of graduate students

BEST: Implementing Career Development Activities for Biomedical Research Trainees
https://doi.org/10.1016/B978-0-12-820759-8.00013-9

and postdocs, our faculty established a course for first-year students on communicating science, and individual departments began hosting periodic professional development workshops within departmental research talk series. In 2009, the Office for Postdoctoral Scholars within the Graduate School of Biomedical Sciences (GSBS) established a monthly Professional Development Seminar Series for both students and postdocs to expose them to scientists in diverse career paths or strategies to enhance their professional skills. These programs were primed to expand when our graduate school created a half-time role in 2012 that was dedicated to the career and professional development of graduate students and postdocs. A few months later, the BEST award mechanism was announced, calling for institutions to develop,

> Innovative new business and academic models of how graduate programs in biomedical research sciences define themselves and their purpose, how they recruit, admit, support, steer, and mentor students to prepare them appropriately for chosen biomedical research or research-related careers [1].

As our campus had minimal career development initiatives, we had the opportunity to design a new approach, starting from a relatively clean slate.

As stated above, we decided to take an approach that was fundamentally different from traditional career development programs by integrating career and professional development into and across all years of the core training experience for all Ph.D. students, rather than placing our primary emphasis on extracurricular, opt-in programs. By integrating career development directly into the curriculum, we sought to even the playing field for all students by making explicit the many facets of professional development and career planning that have traditionally been either under-addressed, addressed to variable degrees within individual mentoring relationships, or defined as "extracurricular," with the undertone that extracurricular activities divert time away from research [2,3]. Importantly, we would be rebranding career and professional development as a core element of training essential for all scientists across all career outcomes — a message that was critical to the culture change we sought to catalyze.

As a group of faculty and staff across the graduate school, we defined learning objectives for distinct areas of professional development, identified points across the pre-qualifying and thesis years at which students naturally use and build on these skills, and designed time-efficient cohort and individual activities to facilitate deeper development of skills and career planning throughout. Building across the full years of Ph.D. training meant that each lesson or activity could build on the last, reducing time investment at any particular stage. Through this just-in-time approach, we taught skills in the context of how the students applied them at their particular stage of training, motivating student engagement.

Since we started, we have been supplementing the core curricular elements with additional opt-in opportunities and services, including student- and postdoc-developed initiatives that we then support and promote, such as site visits to companies via the Industry Exploration Program [4]. We rolled out the curricular changes and programs starting in 2014 via our new Center for Biomedical Career Development (cBCD), which was established within the graduate school as a home for innovation and scholarship in educational approaches for Ph.D. career and professional development. The curricular changes fueled by our BEST award took place alongside other transformative changes to the curriculum designed to address core competencies such as building knowledge frameworks, evaluating information critically, communicating scientific ideas effectively, making evidence-based arguments, and solving problems with diverse teams. This created exciting opportunities for collaboration, synergy, and a more complete integration.

Our graduate school structure lends itself to a curricular approach, with an annual cohort size that gives a critical mass while not being too large ($\sim$50 entering students per year), a single primary umbrella program, and a small physical campus that keeps students within a few proximal buildings. Though students specialize into graduate programs in the second year, they still maintain unified graduation requirements and a single degree (Ph.D. in Biomedical Sciences). In the absence of an umbrella program, our approach could be adapted at other universities via individual graduate programs or departments. And, although we designed each curricular component to build across the full continuum of Ph.D. training, many components could also be adapted as stand-alone elements or to fit within other programs.

In developing this chapter, we chose to describe the curriculum through student voices. We worked together as coauthors encompassing graduate school leadership, cBCD staff, and students, with a particular emphasis on students' firsthand accounts. Though our overall curricular approach addresses a breadth of professional skills competencies, we focus this chapter on the curricular elements that emphasize career awareness and agency in career planning [5].

Planting a seed

A core principle underlying our approach was to encourage students to develop a mindset of ongoing professional development, and we have used early and open discussion of the post-Ph.D. career transition as a motivator for students becoming informed about career options. We plant the seed early — first during recruiting interviews and again during the first week of classes. These first interactions are short engagements with the students: a 15-minute presentation and discussion with prospective students during the interview visit, an incoming-student survey that prompts students to consider their future career interests, and a 30-minute brainstorm and discussion regarding career options for scientists during the first week of classes. We emphasize that there is an exciting and diverse array of career opportunities after the Ph.D. and that part of growing as a scientist means increasing one's awareness of one's own skills and interests to help find one's own dynamic career path.

Third-year student Brent Horowitz, reflecting back on his first year of training:

> During interviews at UMassMed, I heard an emphasis on professionalization and consideration of what we hoped to do with our degrees; this was all before even getting admitted to the program. Once I joined the Ph.D. program, we were asked to consider our careers at the end of Ph.D. training within the first week. At first, I was surprised: I hadn't even picked an advisor yet, let alone decided what I wanted to do half a decade later. However, what the discussion pointed out to me was that I needed to pay attention both to what I value in my career and what my skills are. While I had limited exposure to non-academic careers before coming to UMassMed, in that first class with the cBCD I heard about many potential directions, some of which I had never even considered.

Third-year student Grant Weaver:

> The fact that I'm even thinking about careers at this point in my third year is almost entirely due to the revamped curriculum. I knew before I started grad school that I wanted to pursue a non-academic path but I didn't have any idea what that might look like. We got our first career workshop on our very first day of class. Since then, I've learned about a whole range of career possibilities outside of academia that I was completely unaware of before, and this was completely integrated into my coursework.

Demonstrating our graduate school's commitment to Ph.D. career development starting at the interview stage has had an impact on students' decisions to select UMassMed for their graduate training.

Grant:

> During the presentation at the interview weekend, there was heavy emphasis placed on career development. At the time, this was one of many factors in my decision to attend UMassMed. I remember being impressed with how the graduate school had integrated career exploration and development into the curriculum; but at the time, it was more or less just another mental box checked. Now, as I near the (hopefully) halfway point of my studies, I can see exactly how well I'm set up to find the right career fit for me to transition smoothly into the "real world."

Exposure to career options as a cohort in the third year

Although we introduce themes related to career planning at the onset of graduate training, the strongest career development emphases in the first and second year of training are on professional skills critical to early-stage students' success, such as presentation skills, writing skills, and transitioning into a thesis lab. We chose to return to career planning — with a deeper-dive into career exploration — in the fall of the third year of Ph.D. training. This timing was based on studies demonstrating that students' career interests shift mid-way through training [6,7]. The third year was also chosen because it gives students an opportunity to focus on their scientific coursework, lab rotations, and getting started in their thesis lab before diving deeply as a cohort into career exploration.

In fall of the third year, students take BBS601, a course titled "Professionalism and Research Conduct" (PARC). In this course, we guide students through creating their first Individual Development Plan (IDP) and facilitate discussions on Responsible Conduct of Research. Topics related to the IDP include assessing mentoring gaps; "mentoring up," [8] including strategies for engagement with the thesis committee, research advisor, and career-specific mentors; developing and sustaining a professional network; reflective exercises to consider career interests; strategies for career exploration; developing a professional biography and professional social media presence; prioritizing research- and career-related areas for professional development; and setting goals. As part of this course, students hear from a panel of seven scientists who have pursued a variety of career paths and then have further discussion with two of the panelists in small-group discussions.

Fifth-year student Sumeet Nayak:

During the introductory classes in PARC, one thing that caught my attention was the emphasis on building and having one's own elevator pitch. It was not something that I would have otherwise done or invested my time in, but I soon realized that it is most effective to communicate your work to others in a way that sparks interest in just a matter of a few seconds. The small exercises we did as a class to critique and improve each other's elevator pitch not only gave me confidence to talk about my work but now comes in handy during conferences and many other networking events. I am happy that PARC introduced me to this very concept as I am sure it will help me in any future career I opt for.

Third-year student Heather Loring:

As part of the PARC course, we had the opportunity to meet with a panel, hear about their careers, and ask questions. Through this experience, I discovered potential avenues for me post-graduation and realized I was interested in careers I had not previously considered. As a result, I contacted a connection in one career path I was interested in and learned about a three-day exploration program for current graduate students weighing whether to pursue this career. Without the PARC course, I may not have discovered this opportunity until it was too late to apply.

Brent:

By meeting with current professionals during PARC, I noticed that many of their career paths had been non-linear, winding through a variety of jobs and fields before arriving at their current position. This contradicted advice I had been given before coming to UMassMed, as I had occasionally been told that if someone doesn't find a permanent position in field X within Y years after obtaining their Ph.D., then certain doors would close and not open again. However, several of these professionals had followed quite different paths. I found this reassuring, as I still am not entirely sure what I want to do after graduation, and there's no guarantee my initial choice will be the correct one. I believe that Career Pathways Communities, the next step in the curriculum, will give me additional perspective on how other professionals in my chosen fields reached their current positions.

Diving in deeper via career pathways communities

After considering a variety of career paths in PARC, students take a deeper dive into career exploration via a new model we developed: Career Pathways Communities (CPC), which are learning communities that bring students together to learn about careers of shared interest. Each community meets three times for two hours across a six-week period. Each CPC is themed around one of eight career categories, and is joined by two guest professionals who attend all three meetings and who represent roles within each category.

Meetings are discussion-focused with structured group exercises through which students explore the given career path alongside other trainees with similar career interests, fostering the formation of peer and professional networks. Activities within each meeting are designed to help students compare attributes of various careers related to their community's theme, identify and articulate their own relevant skills and experience, and try a task characteristic of the pathway via a job simulation. At the end of each meeting, students commit to an action they will take prior to the next meeting to continue exploring or preparing for the career path, connecting their CPC experience to their IDPs.

As stated above, the guest professionals participate in all three meetings alongside the students, and they provide advice while addressing questions and concerns the trainees may have about careers within the pathway. Each student participates in two CPCs, one in the spring of their third year and a different one in the fall of their fourth year. This encourages trainees to consider and learn about multiple career options. Postdocs and students beyond their fourth year of graduate training can also apply to participate.

From fourth-year student Daniel Hidalgo:

I participated in the Business & Commercial Development Community during the Career Pathways Communities series. I am interested in pursuing a research career either in academia or industry; however, I was excited to learn more about the business side of science, a branch of science that was an enigma to me.

This community was very informative. The two professionals who joined our discussions were extremely knowledgeable and highly experienced. What made this experience so successful was their willingness to help, as well as their honesty when answering career questions. I believe they understood I was outside my comfort zone; at all times I felt welcomed to understand even the simplest facets of this career. In addition, they pointed me to useful resources (within and outside UMassMed), offered career advice, and provided general information that was helpful in understanding all aspects of the business of science.

The faculty and staff of the BEST program did a great job in focusing their efforts on how to gather and gain the most information possible from the professionals in just three meetings. BEST staff encouraged us at all times to ask questions and set attainable career development actions within meetings in order to be proactive and help us understand this career path better. In-class activities, ranging from informational interviews to job simulations to dissection of job listings, allowed us to capture the true essence of this career. Dissecting job listings — from entry- to high-level positions — with professionals in the field provided a useful insight of the necessary skills and experience (as well as important steps to build or improve on them) for a successful career in business.

As a whole, this experience was fruitful. I gained a vast amount of knowledge and appreciated the importance of business in scientific success; however, highlights of my experience were the contacts made and the professional network built throughout.

From fourth-year student Meghan Spears:

Participating in two Career Pathways Communities has been one of the most valuable contributions to my career development as a graduate student. It allowed me the opportunity to hear firsthand about the experiences of professionals in each career pathway, as well as to ask questions, seek advice, practice relevant skills, and make network connections. After just three meetings for each CPC, I took away a stronger understanding of what it's like to work in each career path and how I can start taking steps now to prepare myself for future success. One experience that stands out most to me is a discussion we had in one of the communities about some of the common fears and misconceptions surrounding careers in that pathway. I found the openness of the discussion and the honest perspectives offered by the professionals to be enlightening and ultimately very influential on my consideration of the career path.

An additional aspect of participating in CPCs that I found to be especially useful in evaluating what type of career might be a good fit for me was the assignment of job simulations…

Trying on career roles using time-efficient job simulation exercises

As the concluding exercise of each Career Pathways Community, students select and complete a job simulation exercise. A job simulation is an activity designed to mimic a task common to the career role. We have developed a library of job simulations, working with professionals representing these roles, that require only one to two hours to complete; as such, we call them #MicroSims [9]. The short time commitment makes #MicroSims an excellent gateway to future experiential learning such as multi-step job simulations [10], externships, or internships. Through the #MicroSims, students "try on" the job role to help assess fit with their individual skills and interests. Students then obtain feedback on their #MicroSim product from the visiting professional in the CPC meeting. This invariably sparks curiosity, leading students to ask a number of additional, deeper questions about the career path, such as how the task fits within the broader role, how it links to other team members' projects, and how one builds skills related to the task. It also gives the professionals a more intimate view of the students' skills, further solidifying the professional relationships being built within these communities.

Meghan:

> Completing job simulation exercises that model tasks done regularly in each career pathway was challenging and thought provoking. However, being challenged in this way allowed me to assess what type of tasks I can see myself taking on in my future career. It forced me to get into the headspace of someone working in that career and to examine what it felt like to do so — something I otherwise may not have experienced as a graduate student. Furthermore, the opportunity to be evaluated on the job simulations by the professionals who are experienced in completing similar tasks was invaluable. Their feedback facilitated the identification of skills that I can improve upon to prepare myself for my future career.

Curriculum integration brings career exploration out of the closet

One of our goals in using a required-attendance, across-curriculum approach was to leverage and even strengthen the peer-to-peer bonds that students form during their first-year scientific coursework. We hypothesized that continuing from scientific coursework into career exploration — as a cohort — would prompt open career-related discussions, facilitate peer-to-peer learning, and bring a sense of community to what can sometimes be an isolating process of career discovery.

Sumeet:

> After hearing professionals on the PARC panel share their success stories in their own respective fields, I wanted to learn and explore more, and that's when the CPC happened. Though I chose "Research in Industry" and "Research in Academia and Government" as my two Career Pathways Communities, I learned a great deal about other career options I was curious about just by interacting with my peers. We would share stories over lunch or coffee from our respective communities. One of these conversations led to a friend introducing me to a professional from one of their communities. I then had an informational interview, which was highly motivating.

Students continue to set their own goals via annual IDPs

After developing their first IDP action plan within the PARC course in the fall of their third year and continuing to set goals within their CPC, students return to the IDP process annually to assess their overall progress, reflect on future career interests, define priority research and professional development goals, and define a one-year action plan. Students are encouraged to frame their goals as "SMART" goals — specific, measurable, action-oriented, realistic, and time-bound — using the framework developed via myIDP [11,12]. Based on input from our BEST Student Advisory Group, we time the IDP assignments to coincide with annual Thesis Research Advisory Committee meetings so that the IDP reflection and action plan process can help students prepare for their thesis committee meeting. The IDP is submitted to cBCD staff for administrative tracking, and students are encouraged to share and discuss their IDP with their advisor, committee, and/or other mentors.

Brent:

> The most important development for my career has been writing and getting feedback on my IDP. Identifying specific skill goals for the year is enormously useful; this gives me actionable items to work on. I find that my long-term and big-picture goals can be nebulous and hard to pursue, but that I can tackle smaller steps toward those goals. Prior to making my IDP, however, I found these difficult to decide on. These SMART goals have been a boon to my career development. For example, becoming a better mentor is a large, undefined goal. Participating in the Peer Mentoring Program [13] and working with college students in the laboratory are more specific actions I've taken toward improving my mentoring abilities. The IDP also helps me be accountable for whether or not I complete my personal goals. Because I can get tunnel vision on my daily or weekly schedule, I have been prone to putting career development to the side. The IDP has been among the most useful tools in my career development.

Heather:

> Having recently completed my first IDP, I can say that the experience is unparalleled in how it helps you delineate your goals, ideas and future plans in a straightforward, digestible format for discussion with mentors. I often intersperse discussion of these big-picture topics during conversations about my day-to-day objectives, but to collect and lay out all my ideas in this format was incredibly fruitful.

Curricular change leading to campus culture change

Building career development into and across the curriculum not only addressed core programmatic learning objectives, but also has sent a clear and recurring message to incoming and current trainees that career development is a core element of Ph.D. training on our campus. As trainees and faculty have embraced this new framing of career development — from being extracurricular activities to being core to students' development as scientists — our campus culture is evolving toward openly valuing the diverse array of careers that Ph.D.s pursue, as well as trainees' preparation for those careers.

Closing thoughts from our student authors

Heather:

These firsthand experiences represent a broader culture that emanates from all facets of the curriculum, priming us graduate students to make knowledgeable decisions about our future. Armed with these opportunities, I have been able to tailor my Ph.D. experience to my career goals instead of blindly rifling through the abundant career options as graduation approaches. The integration of career development throughout the curriculum has instilled confidence that I will be both prepared and informed for what comes next.

Grant:

I was a little apprehensive at first about being a guinea pig [for elements of this new curriculum], but I have been very impressed so far. I am confident that by the time I graduate, I'll know what my next step will be and that I'll have the tools to be successful in whatever profession I ultimately choose.

Sumeet:

Being an international student myself, the two most common and preferred career choices I knew about were either to stay as a postdoc in academia or enter industry as a research scientist, and I myself was bracing for the same. In short, I had a very limited scope in front of me and nowhere to turn. But it all changed after I participated in PARC and the Career Pathways Communities. The mere experience of talking with professionals and UMassMed alumni, especially people who are international and who represented different career pathways, was a great resource to me. It was literally a turning point in my career as a student that suddenly made me realize the many different ways I could use my Ph.D. to channel my career.

Having PARC, followed by CPC, right in the initial years of graduate school has definitely shaped the conversations that I now have with my mentor as well as my peers. As graduate students, apart from our science, we now often discuss the different career opportunities we have after our Ph.D. and what we need to do to make for a better resume. It has made me realize the importance of setting short- and long-term goals and developing the required skill sets and networking that are crucial for pursuing any career path I choose in the future.

Daniel:

At some point during our graduate career we all have questions and fears about the next step, our professional career. It was truly gratifying to frequently hear invited professionals voice their approval of Career Pathways Communities being built into the curriculum, as well as their desire to have had something similar during their time as trainees. Asking for help at times makes us feel vulnerable and is something easier said than done. What is great about UMassMed is that unsolicited career advice comes to us during our graduate training. Transitions from PARC to the Career Pathways Communities feel seamless and flow perfectly with our scientific training and research. This constant support throughout our training years allows us to be well prepared to launch ourselves into a successful scientific career.

Brent:

I felt reassured that I could pursue the career of my choice whether or not my advisor was knowledgeable about that particular path. There will be resources available to me in both traditional and non-traditional pursuits after graduation. If, as I proceed through my graduate training, I find a new interest in a career that my advisor isn't familiar with, I know that I will have the ability to access resources and information about that job.

Meghan:

In my experience, getting students to start thinking about their careers from the very beginning of graduate school and building upon this through each stage of the Ph.D. training creates a mindset that career development should be a continuous buildable process, rather than a last-minute thought as you approach graduation. With the integration of career development into the curriculum from the earliest point in my Ph.D., as a fourth-year student I've already dedicated a significant and beneficial amount of thought to my career options and how I can best prepare for them. I've had in-depth exposure to information, exercises, and discussions both inside and outside of the classroom that have encouraged me to think about my career development, and this has truly motivated me to have a proactive attitude toward my future career.

Closing statement from Dean Lane

Understanding one's own skill set and values, informing oneself of career paths that maximize how skills and values can be utilized and further developed, and taking positive action toward defined goals are essential components for every individual's sense of agency and career actualization. A major focus of our BEST-supported activities has been advocacy for these ideas within the framework of our core curriculum. Highlighting the diversity of opportunities that Ph.D. scientists have successfully pursued, the winding career paths that many professionals travel, and the highly transferable nature of the skills that are central to success in doctoral training contributes to the overall wellness, satisfaction, and ultimately the success of our graduate students. I have enjoyed, as a leader, a team member, and an observer, the process by which our graduate program has developed explicit objectives across these competencies core to career and professional development, and implemented an integrated curriculum aimed at their attainment.

Acknowledgments

Drs. Fuhrmann, Fenn, and Lane acknowledge others who significantly contributed to the development of the career planning components of the curriculum: Anthony Carruthers, Sonia Hall, Anthony Imbalzano, Morgan Thompson, David Weaver, Heather Yonutas, and Phillip Zamore. We also acknowledge the valuable contributions of our BEST Advisory Board and Student Advisory Group, and of the alumni, guest professionals, staff, and faculty who have contributed within the classroom or who supported the development of various components of the curriculum. Curricular innovations reported in this publication were supported by the Office of The Director, National Institutes of Health of the National Institutes of Health under Award Number DP7OD018421 and the Burroughs Wellcome Fund under Award Number 1011612. The content is solely the responsibility of the authors and does not necessarily represent the official views of the National Institutes of Health or Burroughs Wellcome Fund.

References

[1] National Institutes of Health 2013. RFA-RM-12-022. Available from: https://grants.nih.gov/grants/guide/rfa-files/rfa-rm-12-022.html. [Accessed on August 4, 2019].

[2] National Academies of Sciences, Engineering, and Medicine 2018. Graduate STEM Education for the 21st Century. National Academies Press, Washington, D.C.

[3] Leshner AI, Scherer L. Critical steps toward modernizing graduate STEM education. Issues in Science & Technology 2019;35(2):46–9.

[4] Industry exploration program. 2013. Available from: https://www.umassmed.edu/iep/. [Accessed on April 29, 2019].

[5] For further details about our curriculum, implementation approach, lessons learned as we progressed over these five years of development, and data on outcomes and impacts, please see our Educators' Portal, http://BEST.umassmed.edu.

[6] Fuhrmann CN, Halme DG, O'Sullivan PS, Lindstaedt B. Improving graduate education to support a branching career pipeline: recommendations based on a survey of doctoral students in the basic biomedical sciences. CBE-Life Sciences Education 2011;10:239−49.

[7] Sauermann H, Roach M. Science Ph.D. career preferences: levels, changes, and advisor encouragement. PLoS One 2012:e36307.

[8] Lee SP, McGee R, Pfund C, Branchaw J. '"Mentoring up": learning to manage your mentoring relationships'. In: Wright G, editor. The mentoring continuum: from graduate school through Tenure. Syracuse: Graduate School Press of Syracuse University; 2015. p. 133−54.

[9] Center for Biomedical Career Development. 2018. #MicroSim Job Simulation Library. Available from: https://www.umassmed.edu/gsbs/career/educators/microsim-library. [Accessed on April 29, 2019].

[10] InterSECT job simulations. 2018. Available from: https://intersectjobsims.com/. [Accessed on April 25, 2019].

[11] Fuhrmann CN, Hobin JA, Lindstaedt B, Clifford PS. myIDP. 2012. Available from: http://myidp.sciencecareers.org/. [Accessed on April 25, 2019].

[12] Fuhrmann CN, Hobin JA, Clifford PS, Lindstaedt B. Goal-Setting strategies for scientific and career success. Science 2013. Available from: https://doi.org/10.1126/science.caredit.a1300263.

[13] The student-organized Peer Mentoring Program matches third- and fourth-year students with incoming first-year students to provide the first-year students with additional support as they matriculate into Ph.D. training. The program includes brief mentor training.

Implementation of a career cohort model at UNC Chapel Hill: Benefits to students, programs, and institutions

Rebekah L. Layton[1], Patrick D. Brandt[1], Patrick J. Brennwald[2]

[1]Office of Graduate Education, University of North Carolina at Chapel Hill, Chapel Hill, NC, United States; [2]Department of Cell Biology & Physiology, University of North Carolina at Chapel Hill, Chapel Hill, NC, United States

Advantages of the career cohort model

Creating career and professional development opportunities for graduate and postdoctoral scholars can sometimes feel like a daunting task, especially if available institutional funding is limited. Where does one start? What career topics are most important or of most interest? How does one sustain programming on a small-scale budget? These are questions that many, if not all, programs grapple with at some point during their formation. One way to create, maintain, or expand professional development opportunities is to involve trainees in developing programming that meets their needs through career interest groups. This accomplishes a few critical goals, including:

- Empowering the trainees to develop groups related to their specific career interests,
- Providing leadership and networking opportunities to trainees involved in the genesis and administration of the groups,
- Customizing career training options around the areas of greatest need/interest,

BEST: Implementing Career Development Activities for Biomedical Research Trainees
https://doi.org/10.1016/B978-0-12-820759-8.00014-0

- Creating sustainable programming without excessive budgetary requirements, and
- Allowing for easy expansion of programming options without increasing the need for additional staff to support them.

Hence, trainee-led "career cohorts" (career-focused clubs for graduate students and post-docs) are one simple mechanism you can use to create a sustainable on-campus professional development and career exploration program. They are an efficient and cost-effective way to magnify meager resources and maximize trainee success. Although the career cohort model does present some challenges, it provides a solution that can be easily implemented at institutions with no prior experience in providing doctoral-level career programs, or at those wishing to expand into niche career areas. Furthermore, career cohort programming can reduce the activation energy required to get started. A template for implementing such programming is provided in this chapter, along with lessons learned, ideas for how to navigate common challenges, examples of existing cohort programs, and a catalogue of potential benefits.

UNC training initiatives in biological and biomedical sciences

Our professional development branch, known as Training Initiatives in Biological and Biomedical Sciences (TIBBS), serves a population of over 1,000 Ph.D. students and postdocs in the life sciences. TIBBS is embedded in the Office of Graduate Education, which is a cooperatively-funded umbrella program that supports 14 biomedical and biological sciences training programs. The professional development branch of the office was initially developed in 2006 with 1 FTE staff member, although the program has grown to include 2 FTE, Ph.D.-trained directors, and a 0.5 FTE support staff. Nonetheless, programs that are starting with a one-person shop can use the career cohort model to optimize efficiency at the outset.

How are new career cohorts formed?

In the experience of professional development staff members at UNC, the success of career cohorts is based on the influx of student energy and engagement. Participation in career cohorts is voluntary, and trainees can elect to participate in one-time, stand-alone events or series, or can have consistent involvement that can lead to them becoming leaders and helping to plan such events. Importantly, we do *not* create cohorts based solely on staff-perceived needs; rather, we create them when trainees express an interest in participating, volunteering, or practicing their leadership skills. Once two to four key trainee leaders are identified, the career cohort founding process begins. One or more staff members dedicated to mentoring the career cohort leadership team then cultivate and advise the emerging trainee group; a faculty advisor with interests related to the group can also serve as mentor. We have found it advisable to rotate cohort start dates so that the founding of each new cohort is staggered from the others, especially in cases where a single mentor is overseeing multiple career cohorts. In our experience, two new cohorts per year has been the upper limit of manageable cohort creation, as this gives the new cohorts the time to mature, become established, and gain independence. Cohorts at more mature stages require less supervision, allowing newer cohorts to have dedicated mentoring time with the affiliated staff member.

Typically, a staff member dedicates a significant amount of time and effort to train, mentor, and guide club leaders as the club initially forms. This may take the form of biweekly or monthly meetings for four to six months as the club gains experience and, eventually, membership. Initially, the staff member may meet with the new club leaders alone to plan key elements of the club. Establishing a common idea around the club name, mission statement, and purpose can be a great way to start and can help consolidate the interest and energy to drive the next steps of formation. Next, logistical decisions about things like club leadership structures, frequency and format of regular meetings/events, and common meeting times available for future meetings/events should be discussed. Typically, the club leadership starts by creating a shared document (e.g., on Google Drive, Dropbox, or SharePoint) that students can continue to work on and think about remotely after the first few meetings; such a file might discuss bylaws, student club registration requirements, budget/bank account procedures, among other things.

It is crucial to give the trainee leadership guidance around setting a date, planning for, and running the first meeting so the cohort can move on to the next phase. The first event is usually a general interest meeting to publicize the formation of the new club and recruit new members, and the leadership can strategically request that attendees sign an attendance sheet so they can be added to the club's listserv. The interest meeting can be held each year to gauge club interests in various topics/events, but the first one is the most important one, as it helps to confirm and determine the focus of the club; the staff mentor can assist students in brainstorming what information to include and how to structure the meeting.

Club leadership structure is determined by the founding trainee members ahead of time; this could include a chair or co-chairs/co-founders, Executive Board members, or named titles (President, VP, Treasurer, etc.), which enables a divide-and-conquer strategy that reduces the efforts any single person has to make. At the interest meeting, club leadership/officers typically introduce themselves, explain club governance (putting out a call for candidates for roles that still need to be filled), present the drafted mission statement/purpose of the club (asking for feedback in finalizing it if necessary), and present upcoming inaugural club events (three or more events over the course of the following one to three months is a good start).

Annually, the staff mentor should be prepared to orient new club leadership to the roles that they will fill, mentoring them in leadership styles and club structure and training them on software and systems (e.g., room reservations, event registration/attendance tracking, listserv maintenance, advertising event procedures), logistical procedures, and financial procedures/budget (at a minimum). For the first one to two years of a club's existence, it is also beneficial for the staff mentor to meet with the club leadership at least once per semester (or a few times at the start of the semester to get the fall/spring semester programming planned), or regularly every few months to check on programming planning and resolve any issues/questions that may arise.

Challenges

Some common challenges to the career cohort model are the transitory nature of trainee-based leadership, the potential dormancy of clubs if interest wanes, and the challenge of attracting and maintaining trainee leadership. Secondarily, from a staffing perspective, maintaining institutional memory can become an issue as well if staff turnover comes at a time of waning interest in a group. That said, the transitory nature of the trainee-led clubs is also a strength of the model — although some clubs may go dormant for a period of time, the interest will either be renewed by new trainee leadership relatively quickly after the void is felt, or the cohort will not be missed enough to continue programming in that area at the current time. We embrace both outcomes as successful parts of the trainee-led career cohort model, making it incredibly adaptive as new careers of interest become identified and popular. Similarly, the model allows for those same career clubs to ebb and flow in size or even in existence depending on trainee interest or employer need in our local student population.

Standardizing club formation is one way to sustain institutional memory, and creating an umbrella funding structure to support multiple career cohorts under one office or program with distinct branding is a good way to start. The Office of Graduate Education administratively serves that role at UNC, allowing TIBBS to function as the face of professional development while collectively supporting all of the career cohorts across the life sciences departments. Having one faculty or staff member coordinate the umbrella to create brand recognition on campus can legitimize the new trainee-led organizations. Existing student clubs that fill a similar, career-focused role on campus can be invited to affiliate with the funding umbrella, thereby increasing the stability of the umbrella while new clubs form and mature. This approach helps to pilot the career-cohort system and capitalizes on existing interest in other groups on campus. Newer clubs benefit the most, as the umbrella can amplify their impact, reach, status, and visibility across campus, potentially leading to the recruitment of new members.

Starting the funding umbrella/affiliation for the career cohorts often goes hand in hand with providing seed money, which can be as little as a few hundred dollars per year per club to support event refreshments, speaker gifts, student organization fees, website hosting (if applicable), or other material goods such as printing, copies, supplies, etc. Creating a club leader advisory board that meets a few times per year under the umbrella/brand name can facilitate interconnectivity between clubs, leading to communication and coordination necessary for event or speaker co-sponsorship, avoidance of event conflicts, increasing collaboration, reduction of duplicate events, diversification of topics or offerings, and more. Having an additional amount of funds ear-marked for speakers to travel to co-sponsored events can encourage and support club collaborations even further.

The umbrella funding system can also provide student group leadership teams with training on using electronic tracking, event signups, and listservs. Optional additional software training can include website creation (especially if hosting is available on the institutional web platform), social media site creation and maintenance (e.g., LinkedIn groups or company pages, Facebook, Twitter, or Instagram accounts to publicize club events and news), and updates to any existing institutional or program websites. Campus-hosted free web domains, software platforms, and web/media design resources are often available at home institutions and can be recommended to student leaders. These tools can create a common framework for club activities, and listing all of the student groups on a single website as a menu of options can create a powerful sense of institutional dedication to career exploration and development opportunities. Finally, to improve continuity, the umbrella program organizer can also co-administer electronic groups to ensure the smooth transitions between trainee leaders.

Benefits

One of the most appealing aspects of the trainee-led career cohort model is the wide variety of benefits to trainees, program, and institution.

First, **trainees** can gain exposure to multiple career fields as part of the career exploration process. They can learn about their chosen field and gain insight into the culture of organizations and the career itself, the expectations of new hires, jargon, career progression opportunities, and career-specific skill training. Trainee leaders can also gain experience that makes them stand out within a career field by leading field-specific activities, such as case study competitions, or by honing their leadership and presentation skills club through club management. Further, trainees can grow their personal networks to include professionals in their fields of interest who may go on to become professional mentors or potential internship supervisors, or who may simply provide insight into the career path via informational interviews. Personal and professional networks can be crucial in getting referrals, introductions, and job materials passed along to hiring managers during trainees' job search process.

Second, the career development **program** is able to provide a wider variety of programming in both broad and niche areas that likely would not be possible with an equivalent amount of staff time. This model is therefore efficient and effective at multiple levels. Trainees' own personal networks may enhance and diversify the program's professional network, as trainees may invite speakers who come not just from a wide array of career paths, but also from various intersectional economic, social, ideological, racial, religious, and political identities. For instance, many of our career cohorts co-sponsor their events with student organizations focused on affinity groups such as women, LBGTQIA+, Hispanic/Latinx, Native American, African American, first-generation students, international students, and more.

The energy and connections developed by student groups also provide visibility and potential new partners, especially in the local area. To the extent possible, staff should connect students with alumni, professionals, and potential speakers to start the process; however, once the trainees get the hang of reaching out to program alumni and other potential connections through lab graduates, departmental contacts, development offices, or LinkedIn, the contacts they make individually or as a group will exponentially amplify the program's and the institution's networks. As the program develops momentum, it can become a center of gravity that draws in contacts, conversations, and partners from many realms. Robust networking outreach efforts will in turn increase the engagement with recent and future program alumni since these trainees will be familiar with the process and will hopefully be willing to pay it forward to the next generation of scientists by coming back as invited speakers, panelists, or mentors. Connections with alumni just a few years out from graduation are particularly valuable to current students, so career cohorts are an effective way for recent grads to give back, for current students to feel less intimidated in reaching out to new people, and for the program staff to stay engaged with alumni as well.

Increasing the visibility of professional development organizations and/or student clubs on campus can lead to the development of targeted listservs that potential partners can use to offer relevant opportunities, in turn leading to new collaborations or partnerships. Mechanism like these provide structure for clubs or organizations to publicize, enhance, and amplify existing opportunities.

The **institution** can also benefit from program visibility. As the career development program increases engagement with off-campus partners, the campus culture can shift toward acceptance and encouragement of career exploration and professional development of graduate and postgraduate trainees during their training periods. In our experience, this momentum has led young alumni to give back to the program by donating, creating student scholarship funds, and volunteering their time. Finally, by providing robust professional development programs of any kind, institutions can attract a more competitive applicant pool of graduate and postdoctoral trainees, and can increase their own competitiveness for funding sources such as National Institutes of Health NIGMS T32 training grants, among others.

Still not sure where to start?

Here are some career cohorts that we have found enduring and popular, along with their missions:

- **Science Policy and Advocacy Group (SPAG)**, an organized forum where students, postdoctoral fellows, faculty, and staff can learn about and advocate for science policy. Their mission is to promote the role of science in public policy and to expand public access to and awareness of scientific innovation.
- **Science Writing and Communication Club (SWAC)**, formed to foster the intellectual and professional development of aspiring science writers and communicators at all levels of graduate and postdoctoral training. Their goal is to provide club members with career exposure, writing experience, and training opportunities in science writing and communication.
- **Graduate Business and Consulting Club (GBCC)** and **Science and Business Club (SBC)**, focused on exposing current graduate students, professional students, and postdoctoral fellows interested in the junction of business and science to career paths beyond academia, especially careers in consulting.
- **Future Science Educators (FuSE)**, welcoming graduate students and postdocs interested in science education, administration, and science outreach. Their focus is to develop job application materials and career exposure, and to build skills/experience relevant to teaching-intensive careers.

- **Academic and Research Intensive Careers (ARIC)**, preparing students and postdoctoral fellows for a career as principal investigators in research-intensive environments by developing the skill set required to be competitive in the current job market.

Logistically, once the three- to four-person critical mass of trainee leadership has been reached, an active club membership of at least 15–20 members (8–10 attendees per event) is ideal. Some clubs have closer to around 80 active members (regularly garnering at least 20–30 attendees per event), so the size can vary quite a bit. Continued monitoring of club interest and attendance/participation (including through annual surveys) can help determine the health of a club, continued funding, etc.

Workshop series

Career skill-building workshop series can broaden the impact of single events by linking their theme, career path, or developed skill into a larger set of events. This often yields higher attendance per event (30–50 per event at UNC) and affords a deeper dive into a particular career pathway since workshops can build upon previous experiences. We offer two to three workshop series per year and rotate them every few years. The two most popular series, covering industry skills and teaching, are offered every other year; specialty series are brought back every two to three years. This staggering of series allows for broad career exploration, higher visibility for specific careers associated with higher trainee interest, and skill and knowledge building in various areas necessary for trainees to stand out as exceptional job candidates. TIBBS co-sponsors workshop series with UNC clubs or with external partners (such as local universities) to help spread the experience, which benefits students who can network across institutions and helps programs, clubs, and institutions by splitting the cost of expensive events or speakers. Sample series are presented in detail below, and a sample schedule of events and speakers for the workshop can be found in the appendix.

Academic and research-intensive career series

This is a seven-part series targeting research-intensive career exploration and professional development, especially for trainees interested in becoming principal investigators or team leads in academic, government, or industry labs. Session topics include understanding the academic career path and its many variations; successfully preparing for and navigating the job market; making the most of a postdoctoral experience; challenges of running a lab from early-career stage investigators; principles of effective mentorship; differences (and similarities!) between academic, industry, and government research-intensive careers; networking skills and tips; and navigating lab budgets and funding. Sessions typically specialize in a particular career pathway branch (academic, government, industry) for a deeper dive, or touch on all three branches.

Leadership series

Presented as a six-part workshop series aimed at empowering biomedical trainees to discover and develop their own leadership potential, this series offers an opportunity to learn more about trainees' own strengths and weaknesses, engage in structured and meaningful professional development activities, gain a deeper understanding of what makes them "tick" as scientists and as future leaders, and develop their confidence and competence as professionals. Participation allows trainees to both develop and demonstrate leadership, management, and interpersonal skills that improve performance as students and as future employees, all while interacting with students from Duke University through this co-sponsored workshop series.

Teaching series

Presented as an eight-session series consisting of pedagogical training for best practices in teaching, it includes topics such as interactive team learning; learning assessments; Process Oriented Guided Inquiry Learning (POGIL); developing a teaching statement, portfolio, and application materials; classroom management; syllabus design; challenges of first-time teachers; and active learning. This workshop will be co-sponsored by FuSE in the future.

Science policy series

This series consists of seven seminars and four workshops that develop trainees' skills and experience in science policy and advocacy by covering topics such as written policy assignments, science policy careers and skills, informal science communication, communicating science to policymakers, leveraging STEM investments for maximum impact, technology transfer, advocacy, and the science of science policy. This series is co-sponsored by SPAG.

Science communication series

Presented as a six-session series, this series offers two workshops where trainees can submit written content for peer review to help them develop their communication skills. Other topics include science journalism, presenting science to the public, editing and evaluating scientific publications, ethics in science writing and communication, and science blogging and social media. This workshop is co-sponsored by SWAC.

Industry skills series

Presented as a seven-session series that exposes trainees to professional knowledge and skills crucial to industry careers through interactive sessions, this workshop series covers topics such as: leadership, teamwork, and professionalism; industry culture; personality and leadership style; interviewing skills; networking; and industry career options. A co-sponsorship by GBCC and ARIC is planned for the future.

Travel funding

If additional resources are available, providing competitive travel funding is an additional way to maximize students' experiences and opportunities through the career cohort system. At UNC, we have deployed various models for funding trainee travel. One way of establishing a competitive travel award is to put out a call for documents during each of the workshop series and select the top peer-rated document using a standardized, student-developed rubric; the documents can be relevant to developing career-related skills. Sample documents include: teaching portfolios, teaching-focused CVs, teaching-as-research projects, and draft assessments for the Teaching Series; reflection essays on leadership/communications styles following participation in 360-assessments (a form of feedback for leaders from all directions - supervisor, peers, and direct reports/mentees, if applicable) and an experiential leadership high- and low-ropes course challenge for the Leadership Series; and peer-reviewed science policy memos and white papers for the Science Policy Series.

Another model for distributing funds is through a competitive application process. This option simply requires the creation and development of an impartial application and judging process (e.g., submission of statement of interest, CV/resume, proposed budget, or other materials of interest). Alternatively, you can distribute travel funds to career cohorts for their use on group trips related to the career field; for instance, GBCC has used travel funds to travel to case competitions. Using travel awards is a very flexible option that can be adjusted as needed. A sample travel award application for FuSE and a flyer describing how to obtain a travel award during a Science Policy workshop series are included in the appendix.

Conclusions

The trainee-led career cohort model is one way to jump start and organize career and professional development events for newly established or under-resourced programs, as it enables robust programming even in cases in which resource or staff support is limited. It can even serve as a way for established programs to broaden the availability of career exploration and professional development, especially for niche or emerging career areas that may otherwise be overlooked. We believe this model affords enormous opportunities for multilayered benefits that can easily overcome many challenges and potential concerns. Trainees benefit from opportunities for leadership experience, from network expansion, and from the career exploration and training itself. When they lead the programming, trainees are more likely to get exactly what they want and they need.

Acknowledgments

This work and these programs were made possible by an NIH-BEST award to the University of North Carolina, Chapel Hill (DP7OD020317). The FuSE Travel Award application (Appendix 2) was developed by Future Science Educator (FuSE) club leaders, including Brittany Miller, Aspen Gutsell, Melinda Grosser, Melissa Babilonia-Rosa, and Kelsey Gray. The Science Policy Series Certificate flyer (Appendix 3) was developed by Science Policy and Advocacy Group (SPAG) club leaders and the Science Policy Series Planning Committee. Thanks to Jessica Griswold, who designed the poster, and to Edhriz Siraliev-Perez, Jennifer Kernan, Kelsey Miller, and Yael-Natalie Escobar.

Appendix 1. Sample club events schedule

UNC TIBBS & ImPACT
Training Initiatives in Biological and Biomedical Sciences
Immersion Program to Advance Career Training
Select Events from 2017-2018

September 2017	Co-Sponsor
Seminar on College Teaching with Ed Neal	GRAD/SPIRE/FuSE
NSF Graduate Fellowship funding workshop	
Science Policy Workshop Series	SPAG
Pfizer Information Session	
Medical Science Liaison Career Talk with Wendy Toler	MSL@UNC
F31 Grant Writing Workshop	
Author/Editor Orientation and Training	SWAC
INC Research Site Visit	ELITE
October 2017	SPAG
Science Policy Workshop Series (cont'd)	SPAG
Medical Science Liaison Career Talk with Chris Brodie	MSL@UNC
Academic Career Pathway Panels – Faculty and Administrative Careers	GRAD/FuSE/ARIC
The Journey to the Top – Women's Paths to Success with Jessica Polka	WinS
8th Annual Career Blitz	
Mentoring Doctoral Students for Non-Academic Careers with Dr. Paula Chambers	GRAD
Beyond the Bench with Tim Martin	SBC
Center for Open Science Workshop	
November 2017	
Career Networking Lunch with Jessica Polka on Policy/Advocacy	SPAG
Learning to Network and Interview Successfully	GRAD
Science Policy Workshop Series (cont'd)	SPAG
Medical Science Liaison Career Talk with Shay Taylor	MSL@UNC
M.S./Ph.D. career Fair	UCS/GRAD
Career Networking Coffee/Cookies with Dr. Galloway, Dr. Davison, and Mr. Corb	SPAG
Beyond the Bench with Lauren Neighbours, Clinical Research Scientist at Rho	SBC
The Journey to the Top – Women's Paths to Success: Crystal Harden	WinS
Beyond the Bench with Michael Dial – Principal at Hatteras Venture Partners	SBC
Consulting Bootcamp Weekend Retreat	SBC/TIG
Omnicom Health Group PostDoc InfoSession	

December 2017	
ImPACT internship Showcase	
The Journey to the Top—Women's Paths to Success: Meg Powell	WinS
Mindfulness in the Lab: Reduce Stress, Improve Your Focus!	
Education Research Alumni Seminar by Melissa Babilonia-Rosa	FuSE
Medical Science Liaison Career Talk with Chad McKee	MSL@UNC
1—1 career counseling with Denise Saunders	
January 2018	
ImPACT Info Session	
Practical Skills Workshop Series: Resumes for Industry	WinS
Science Policy Advocacy with Francis Colon	SPAG/SACNAS
Career Networking Breakfast with Francis Colon	SPAG
IMSD Career Planning Workshop	IMSD
Business Essentials Mini-Course with Brian Buxton	
February 2018	
Career Planning Workshop	
HHMI Pedagogy Workshop	GRAD
1—1 Career Counseling with Denise Saunders	
ClearView Info Session	
GenScript Biotech Info Session	GEN
BD Info Session	BD
IMSD Advanced Career Planning Workshop	IMSD
F31 Grant Writing Workshop	
Practical Skills Series: Overcoming Imposter Syndrome Workshop with Valerie Ashby	WinS
Beyond the Bench with Ryan Hallett and Tim Jacobs	SBC
Science Communications Series	SWAC
March 2018	
Advanced Career Planning Workshop	
Dissertation Formatting Workshops	GRAD
GenScript Information Session/Antibody Drug Development Workshop	
Career Networking Coffee/Cookies – Clinical Pharmacology	AAPS
LinkedIn Workshop	
Practical Skills Series: Salary negotiation with Claire Counihan	WinS
Diversity in STEM Conference	IMSD/CSS
Coffee & Conversation with Marsha Massey	FuSE
Business Career for Life Science PhD Panel with Brian Buxton	SBC
ComSciCon Triangle	SWAC
Science Communication Workshop Series	SWAC
Intellectual Property Rights Workshop	SBC/SPAG

April 2018	
Postdoc Boot Camp	ARIC
Beyond the Bench with Drew Appelfield	SBC
Women in Science Symposium	WinS
Career Networking Coffee & Breakfast with Lydia Villa-Kamaroff	ARIC
Carolina Innovations Seminar	
Jackson Laboratories Information Session/Perfecting your pitch application workshop	JAX
Science Communications Series	SWAC
Career Networking Coffee/Cookies with Pamela Felicitano	ARIC
May 2018	
ARIC Spring Meeting	ARIC
College Science Teaching Info Session	GRAD
Finance Workshop with Andrew Hardaway	FS2
June 2018	
Career Options for STEM PhD Researchers @ NCBiotech	NCBiotech/EXT
Genscript Seminar	GEN
F31 Workshop	
Financial Sense Meeting with Andrew Hardaway	FS2
IQVIA – Novella Info Session	NOV
TIBBS Teaching Series	FuSE
July 2018	
TIBBS Teaching Series & CIRTL Short Course	FuSE
August 2018	
TIBBS Teaching Series	FuSE
TIBBS Welcome Back Picnic	

Student group co-sponsor abbreviations:

SBC = Science and Business Club, **SPAG** = Science, Policy and Advocacy Group, **SWAC** = Science Writing and Communication, **FUSE** = Future Science Educators, **WinS** = Women in Science, **ARIC** = Academic and Research Intensive Careers, **MSL@UNC** = Medical Science Liaison group; **SACNAS** = Society for Advancement of Chicanos/Hispanics and Native Americans in Science; **AAPS** = American Association of Pharmaceutical Scientists; **FS2** = Financial Sense for Scientists.

Organizational/Departmental/External co-sponsor abbreviations:

OPA = Office of Postdoctoral Affairs, **GRAD** = the Graduate School, **UCS** = University Career Services, **SPIRE** = Seeding Postdoctoral Innovators in Research & Education, **ELITE** = Enhancing Local Industry Transitions through Exploration – UNC/Duke/NIEHS Consortium; **IMSD** = Initiative for Maximizing Student Development; **CSS** = Chancellor's Science Scholars Program; **BD** = Beckton-Dickinson; **JAX** = Jackson Laboratories; **NOV** = Novella; **GEN** = GenScript; **TIG** = Triangle Insights Group.

Example of the UNC TIBBS 2017–18 programming calendar including career cohort programming.

Appendix 2. Sample instructions for club travel award application

Frequently asked questions and selection criteria for travel funds awardees

Who will review the application?
The FuSE board will review the application and make a joint decision. If a board member applies, that individual will be excluded from voting on that round of applications.

How often will the travel awards be offered?
We expect to offer at least 2 awards for each of the Fall and Spring semesters, but this will depend on the registration fee and travel costs of the people selected each semester. If there is any remaining budget, a summer application can be opened or the money can be added to the pot for the next semester.

How much money will a participant be awarded?
The amount of award money will be determined on a case-by-case basis. The decision will be dependent on the registration fee for the event as well as the itemized travel expenses. The maximum amount that will be awarded per semester is $300. For proposed trips totaling more than $300, participants should note whether other funding sources will help cover expenses or whether the participant is willing to cover the rest of the expenses.

Where can I find out about conferences to apply for?
Join the FuSE Listserve to get more information about upcoming conferences as we learn about them. To research on your own, the Elon Center for the Advancement of Teaching and Learning has a database of recommended teaching conferences
(http://www.elon.edu/e-web/academics/teaching/conferences.xhtml) as does the Society for the Advancement of Biology Education Research
(https://saber-biologyeducationresearch.wikispaces.com/Meetings+of+Interest+to+Members).

Selection criteria
Applications will be evaluated by the following criteria (for a possible total score of 20 points):
1. What are the career goals of the participant? Are the career goals in alignment with attendance to the proposed conference/event? (1- event not well aligned, 5- event very well aligned)
2. How will the participant share with their FuSE fellows the teaching techniques or pedagogical content learned in the conference/event? How valuable will the proposed information/materials be to club members? (1- not at all valuable, 5- very valuable)
3. Participants with previous involvement in science education or science outreach activities will be prioritized. (1- poor/no previous involvement, 5- excellent/a great deal of previous involvement)
4. Is the proposed budget reasonable/realistic? (1- not at all reasonable, 5- very reasonable)

For graduate students: priority will be given to students who have completed their oral and written examinations as well as students that are closer to graduation. Please include this information in your application materials.

FuSE Travel Award Application:

FuSE FUTURE SCIENCE EDUCATORS

Applicant's Name:
Select (fill) Graduate student (___ year) or PostDoc
Email Address:

*** Please include a copy of your CV (attach separately)**

Conference Information:
Name of Conference:
Date of Conference:
Address of Conference:

What is the distance (in miles) to your event?

Please provide a total estimate of travel costs below with an itemized estimate of costs that you would like considered for coverage (e.g., conference registration fee, hotel, transportation, etc.)
Conference or Registration Fee:
Hotel/Lodging:
Transportation:
Other (specify):
TOTAL COST TO ATTEND (est)

How will your participation in this event/conference benefit your career goals? (300 words max)

List what deliverables you will bring back to your FuSE fellows if you attend this event/conference. How will this benefit other club members?

Electronic Signature of Applicant:
Date:

Example of a career-cohort travel award application developed by Future Science Educators (FuSE).

Appendix 3. Sample flyer for a workshop series, including requirements for the travel award

Example of a workshop series flyer developed by the Science Policy and Advocacy Group (SPAG) and the Science Policy Series Planning Committee, including series participation requirements.

Vanderbilt's ASPIRE program: Building on a strong career development foundation to change the Ph.D.-training culture

Kimberly A. Petrie, Ashley E. Brady, Kate F.Z. Stuart, Abigail M. Brown, Kathleen L. Gould

Biomedical Research and Education Office, Vanderbilt University, Nashville, TN, United States

Introduction

Context

The Vanderbilt University School of Medicine (VUSM) has made major commitments to graduate student and postdoctoral training for many years. In 1998, VUSM established the Office for Biomedical Research, Education and Training (BRET) with the goal of providing integrated support for research trainees. In 1999, BRET established one of the first Offices of Postdoctoral Affairs in the country, and in 2005, it established an Office of Career Development and an Office for Outcomes Research. In 2011, additional staff and leadership charged with expanding career development initiatives were added to the BRET Office. The ASPIRE program was therefore created in a highly supportive institutional environment and was built on a strong core foundation of providing career and professional development activities to biomedical sciences Ph.D. students and postdoctoral scholars. The ASPIRE program is now supported by four full-time and one half-time faculty and staff.

Goal

The goal of the Vanderbilt ASPIRE program was, and continues to be, to empower and prepare biomedical sciences Ph.D. students and postdoctoral scholars (collectively called

BEST: Implementing Career Development Activities for Biomedical Research Trainees
https://doi.org/10.1016/B978-0-12-820759-8.00015-2

trainees) to make well-informed career decisions, and to broaden their experiences so that they transition efficiently to research and research-related careers in non-academic and academic venues. We achieve this through multiple mechanisms, including providing a variety of professional development, career exploration, and training enhancement activities in which trainees can participate at stage-appropriate times during training (Fig. 15.1). We engage external and internal partners to offer educational and experiential training opportunities, including non-didactic modules in business and entrepreneurship, teaching, communication, data science, and clinical research, and externship and internship opportunities for interested trainees to gain deeper practical exposure to a career area of interest. Together, these ASPIRE programs shift the focus from preparing trainees solely for productive academic research careers to preparing them for productive careers in the biomedical research enterprise, broadly defined, that of course includes a career as a tenure-track faculty member at an R1 institution.

The ASPIRE program offers a new paradigm for biomedical Ph.D. and postdoctoral training at Vanderbilt University and integrates career and professional development into the training experience. ASPIRE is designed to enhance the career readiness and resilience of our trainees without lengthening time-to-degree for Ph.D. students or time-to-completion for postdocs.

Scope

The development of each ASPIRE initiative was profoundly influenced by the results of surveys of then-current trainees' career interests and ongoing analyses of alumni careers. Not surprisingly, we found that there is tremendous diversity both in the career interests of our current trainees and in the actual career outcomes of our alumni. We therefore developed the ASPIRE program so that all trainees could identify components of value to them and participate no matter their career interests. ASPIRE activities cover professional skill-building, career exploration, and experiential learning that is tailored to trainees at different stages of training. Some activities are targeted towards Ph.D. students only and others are offered for postdocs only; the experiential learning opportunities are available to both populations of trainees. Except for introductory elements of the program for first-year Ph.D. students, all ASPIRE offerings are voluntary.

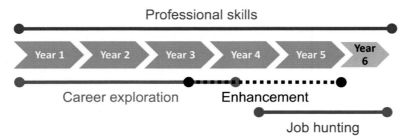

FIG. 15.1   Intentional career planning for Ph.D. students by year in training.

Building ASPIRE

Progressive construction

We developed and implemented the various components of ASPIRE in yearly increments that matched the progression of the cohort of Ph.D. students that began their training when ASPIRE launched (Fig. 15.1). Thus, the full suite of proposed activities was fully implemented only in the final year of NIH BEST funding.

In the first year of BEST funding, we gathered together and created career and professional skill development resources relevant to first-year students and shared these materials (slide sets, reading lists, and interactive exercises) with the faculty leaders of IMPACT groups. IMPACT (Intensive Mentoring Program for Advancement and Career Training) is a required, non-credit-bearing course designed to ease the transition to graduate school. In IMPACT, small groups of 10–12 first-year students meet with a faculty mentor for one hour each week throughout their first year. The IMPACT toolkits we created are used by faculty mentors to facilitate discussions on topics relating to professionalism and career planning.

Also in the first year of BEST funding, we began developing career exploration resources and activities appropriate for second-year students and postdoctoral fellows. These included the *Beyond the Lab* video series [1] for career exploration and the *ASPIRE to Connect* [2] workshop that introduces networking strategies. We refreshed our existing *Annual Career Symposium* [3] event and expanded our *Ph.D. Career Connections* [4] seminar series. Lastly, we introduced a non-didactic, short course in career planning and began our externship program, defined as one- to three-day job shadowing experiences. A few externships are facilitated by our office (e.g., a science policy/advocacy trip to Washington, DC, with Vanderbilt's Office of Federal Relations) while others are organized by individual trainees.

In the second year of NIH BEST funding, we developed the *ASPIRE Café* for postdoctoral fellows. Postdoctoral fellows often report feeling a lack of community and a dearth of organized institutional support during their training. This series, which is described in more detail in following sections, caters to the specific professional development needs of postdoctoral fellows. In Year 1, we also launched six ASPIRE modules: non-didactic short courses covering topics in communication, teaching, clinical research, data science, and business and entrepreneurship, often led by campus partners. Two examples are the partnership with the Center for Technology Transfer and Commercialization (CTTC), which provided an eight-week module that introduced technology transfer principles, and the partnership with the Assistant Director of VUMC News & Communication, who teaches a seven-week module on biomedical research and the media.

In Year 3, we expanded the module program and launched the ASPIRE internship program. Internships are typically paid, part-time (8–10 hours per week), hands-on experiences in which a trainee tackles an in-depth project over the course of 10–12 weeks. Most ASPIRE internships are at sites in the Nashville area, or are performed remotely through partnerships with national organizations. We developed new internship opportunities *de novo* through extensive relationship-building efforts and outreach to prospective employers, local companies, and organizations, and also assembled a list of established internship programs in large companies to which trainees could apply. That year, we also added smaller initiatives such as "Lab to Lunch," a dining etiquette overview event that improves trainee confidence for meals associated with interviews and conferences, and "Headshot Day," which provides professional photography services to generate job-ready portraits.

In Years 4 and 5, we continued to offer all existing programming and to build on the ASPIRE module and internship programs. By the end of Year 5, the final year of NIH funding, we had developed a suite of 11 modules (Table 15.1), had created ~30 internship opportunities with remote or local host organizations, and developed *ASPIRE on the Road*, a company site visit initiative. *ASPIRE on the Road* trips leverage alumni contacts to arrange site visits at industry or academic venues across the country, and provide trainees the opportunity to learn firsthand about different career options from professionals in a range of roles.

To develop, implement, and continue to carry out ASPIRE activities and events, the efforts of two education-track faculty, two staff, and one part-time research faculty member are required. These efforts are complemented by those of one part-time evaluator who analyzes the impact of these activities on trainee outcomes.

Broadened online and in-person communication strategies

As part of our initial efforts to build ASPIRE, we constructed a robust website [5] to house information about our programming. This provided us with a one-stop-shop to advertise events, highlight resources, share our professionalism policy, and explain who we are and how students, postdocs, alumni, and employers can engage with us. The website also hosts an active blog that highlights current activities and resources, and that archives past opportunities and events that are easily searchable through a powerful tagging system.

Beyond establishing our website, we branched out in different directions to reach our trainees through social media. We now host two Twitter accounts — one exclusively for job postings [6] and a separate account that we use to highlight our events and resources [7]. Our YouTube alumni interview series, *Beyond the Lab*, informs our students and postdocs (as well as non-Vanderbilt trainees) of career opportunities and strategies while also celebrating alumni achievements. Indeed, the YouTube series was so popular and foundational (our videos have collectively accrued over 31,000 views) that we launched a companion podcast series where trainees can easily listen to the advice contained in the YouTube video series through an audio-only format.

TABLE 15.1 ASPIRE modules and when they are offered each year.

Fall	Spring
Introduction to principles & practice of clinical research	
Practical strategies for strong writing	Practical strategies for strong writing
Technology commercialization	Management & business principles for scientists
Creating effective scientific talks	Clinical laboratory medicine:applying your Ph.D. to patient care
Biomedical research & media	K-12 STEM education
Networking pacing	creating effective scientific talks
EQ + IQ = career success	Data science essentials

We also began a campaign to reach out to alumni, parents of current students, and employer partners to share news about our trainees and our office initiatives. The *Results and Discussion* newsletter is a key component of this strategy. The twice yearly, full-color glossy newsletter highlights trainees and their research achievements, interesting alumni and faculty members, and unique career development programs. The newsletter is written by current trainees (many of whom have participated in our *Biomedical Research and the Media* module), as it provides them with the opportunity to practice writing and editing for a lay audience. Recipients of the newsletter are pleased to receive these periodic updates about the goings-on in the BRET Office and the achievements of their fellow classmates, their children, or, simply, the trainees of an institution they support.

Fostering faculty buy-in

We used several strategies to educate and engage faculty in the development and implementation of ASPIRE. Faculty engagement began during the first year of our program when ASPIRE PIs presented the findings and recommendations of the 2012 NIH Biomedical Workforce Working Group at more than 15 faculty meetings throughout Vanderbilt's biomedical enterprise. Included in these presentations were data about the career outcomes of Vanderbilt biomedical graduate students and how these data relate to national career outcomes data provided in the NIH workforce report. Vanderbilt outcomes data were further broken down so that faculty could appreciate the career outcomes of students from their departments. Then, the PIs presented the goals and strategies for developing ASPIRE over a five-year period and invited the faculty to engage with the program. As a result, a subsequent survey of the faculty in 2016 revealed that 80% of the them were aware of the ASPIRE program and were supportive of its mission [8].

In Years 2–5 of the program, the ASPIRE PIs also gave presentations at multiple standing meetings of directors of graduate study, department chairs, executive faculty, and junior faculty development groups, as well as at all-encompassing School of Medicine faculty meetings. A goal of these presentations was to describe the components of the ASPIRE program as they were developed, and to describe which activities were appropriate for each stage of student and postdoc training. A second goal was to seek faculty perspective about how the program was working and to assure the faculty that our vision was to partner with them in the training of their students and postdocs, not derail trainees from their research. A third goal was to seek faculty engagement, including participation on ASPIRE faculty panels, in modules, and in professional development sessions. As a result of these efforts, faculty members now routinely contact us when they learn of an alumnus returning to campus who can share information with current trainees, and frequently share career opportunities that may be of interest to current trainees.

In Years 4–5 of the program, ASPIRE leadership met one-on-one with every training grant director and every director of graduate study in the biomedical enterprise to share the breadth of ASPIRE career and professional development sessions that their trainees could attend. These training leaders, together with their trainees, now select from a menu of activities and programs tailored to their interests and needs. Training program leaders suggested additional professional development session topics and ASPIRE staff and faculty then developed these resources designed to meet the present needs of each training group. These new sessions include "Maximizing Your Research Potential" and "The Next Step: Applying for Your Post-grad Job or Postdoc."

Finally, as we upload each *Beyond the Lab* video, we contact the former faculty mentor of each alumnus and provide them with a link. The faculty are overwhelmingly pleased and proud of their former trainees' accomplishments and performance in the interviews.

Identifying and developing relationships with campus, local, and national partners

Partnerships have been essential to the development of all of our ASPIRE program activities. Our first partnerships were developed to serve our module program, tapping campus partners like the VU CTTC and the VU Writing Studio as experts to teach topics not covered in disciplinary coursework. Early in the grant, we also began reaching out to local life sciences companies and non-profit organizations to develop opportunities for VU trainees to broaden their experiences outside the laboratory. Not surprisingly, relationship-building requires a significant time commitment and patience to be able to reap the benefits of these partnerships. Thus, at the inception of our BEST grant, we expanded our team to include a Director of Strategic Partnerships who would spend a significant portion of effort building relationships with employers and alumni. VU is located in Nashville, where the life sciences industry is fairly small. In some respects, this has made involving local partners in the building of our programs more challenging; however, this comes with the benefit of relatively easy access to Tennessee companies who are eager to build a pipeline of local talent.

To identify potential partners, we have attended and spoken at conferences, venture capital pitch competitions, and networking events targeting the local life sciences and entrepreneurial ecosystem. Once identified, potential partners can engage with our program in multiple ways, including through meetings with our team, assistance with recruiting new hires by networking with our trainees, coordinating a site visit at their organization, or hosting an intern. We have found a willing and eager partner in our local life sciences industry advocacy organization, Life Science Tennessee, and its many member organizations. We have also relied heavily on our ASPIRE Advisory Committee, which meets annually and is made up of individuals strategically chosen from a range of industries who are well positioned at a national level to advise us and to connect us with others; it should be noted that a graduate student and a postdoc representative also sit on the Advisory Committee.

It has been important to recognize that partners have different motivations for working with us. For example, alumni want to give back, companies want access to smart individuals for their talent pipeline, non-profits need more manpower to support their mission and develop their own pipeline, or professionals simply enjoy the opportunity to mentor and interact with energetic trainees. The key to successful partnerships is to identify their motivations early and to gauge the degree to which partners want to be involved — a factor that can change over time.

The six facets of the ASPIRE program

Built on strong institutional commitment and a solid foundation of career exploration programming and advising delivered by a single career development professional, ASPIRE expanded the reach, vision, and staff dedicated to supporting trainees' professional development. ASPIRE activities can now be placed in one of six categories or facets (Fig. 15.2). These interlinked facets allow us to fulfill our mission of helping prepare Ph.D. and postdoctoral trainees for their next career stage.

Career exploration and decision-making

Providing opportunities to explore a breadth of careers is paramount to the ASPIRE program. Our largest event in this arena is our *Annual Career Symposium* that rotates its primary focus around the topics of academic, industry, communication, science policy, and health-related careers. Keynote presentations, panels, workshops, and table discussions are led by selected alumni and ASPIRE Advisory Committee members, with a networking reception capping off the day. Pre-registration has ranged from 325 to 475 trainees over the past five years. A dinner the day before the event allows our alumni speakers to reengage with former peers, mentors, and members of the ASPIRE team.

Ph.D. Career Connections is a monthly one-hour seminar series featuring an alumnus discussing their career path and current position. Each speaker is given a prearranged set of questions to address in the first 30 minutes so that current trainees have plenty of time for questions afterwards. In addition to these seminars, we host a variety of alumni and employer sessions in a similar format, and also organize panel presentations for current trainees to describe various externship and internship experiences to their peers. These in-person sessions are complemented by the *Beyond the Lab* video and podcast series.

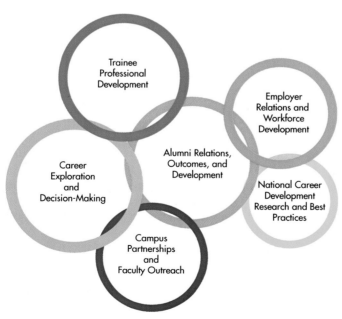

FIG. 15.2 The six facets of ASPIRE.

In addition, trainees are encouraged to design individual externship experiences that we currently support with travel awards. As examples, trainees have participated in the AAAS Catalyzing Advocacy in Science and Engineering workshop in Washington, DC; attended Biogen's Drug-Discovery Conference in Boston, MA; attended the conference of the Defense Threat Reduction Agency; and traveled to the American Medical Writer's Association annual conference in Orlando, FL. In addition, every other year, ASPIRE joins efforts with the Vanderbilt Office of Federal Relations to take trainees on a STEM policy externship trip to Washington, DC.

Finally, individual advising remains a cornerstone of our approach with ~150 career advising appointments per year. Discussions focus on professional development or job search activities and include preparing and applying for academic and postdoc positions; pursuing opportunities outside the academy; preparing CVs, resumes, cover letters, and biosketches; honing presentation skills; practicing interviewing and negotiation; and reviewing graduate school survival skills.

Professional development

Trainees need to develop a variety of skills in addition to research proficiency for any career they choose to pursue. We offer professional development activities designed to complement and integrate with the training provided by graduate programs and research advisors.

Each week, our office hosts an open hour to review CVs, resumes, biosketches, and cover letters from trainees who bring prepared documents. This popular activity brings over 100 trainees to the office each year.

ASPIRE to Connect is an annual, half-day workshop that offers practical tips for meeting new people and cultivating authentic connections. It typically features an expert in interpersonal communication whose talk is then complemented by breakout sessions on topics such as identifying professional networking opportunities, video interviewing, telling the story of your research, leveraging LinkedIn, and making the most of a professional conference. While this stand-alone event has attracted 60–140 trainees each year, we intend to either offer it every other year to attract more trainees at each occurrence or incorporate the content into our *Annual Career Symposium*.

The ASPIRE modules (Table 15.1) are short, non-credit bearing electives that broaden the training experience. They are intended to be minimally disruptive to trainee schedules by requiring a limited time commitment on a weekly basis (one to six hours) over a protracted period of 6–12 weeks. Some modules are designed with enrollment caps to facilitate personalized training while others are open to any interested trainee. Most require that student participants have passed their qualifying exam. To date, 676 trainees have participated in the ASPIRE modules, and anecdotally, numerous trainees have cited their module participation as an important factor in their successful job search.

To ensure that postdoctoral fellows have access to professional skills training and advising pertinent to their unique needs, we launched the *ASPIRE Postdoctoral Café* (or simply, *ASPIRE Café*) early in our program. Now, it is an established, twice-monthly seminar series. We offer a

TABLE 15.2 Representative *ASPIRE Postdoctoral Café* session topics.

Sample sessions
Recognizing and overcoming the imposter syndrome
Behind the curtain: an inside look at peer review of F32 NRSAs
NIH career development (K) awards: Which one is right for you?
Grant writing resources
Preparing a faculty application package
Science and social media
Lean lab management
What to expect for a faculty position interview
How to give a chalk talk
Negotiating your faculty start-up package
Mastering the art of interviewing
Informational interviewing
Exploring and preparing for faculty careers
A fair look: How to review a paper
How to give an effective scientific talk

quarterly *Orientation to Career Development Resources* session for newly on-boarded postdocs. The remaining sessions cover a variety of professionalism and career development topics (Table 15.2), including topics pertinent to academic career trajectories that are specifically tailored to the needs of postdocs. *ASPIRE Café* sessions are often led by faculty or other campus partners.

Employer relations and workforce development

Developing strategic partnerships with various companies and organizations at both the local and national level is critical to the continued growth of ASPIRE. Throughout the year, we have conversations with employers, the ASPIRE Advisory Committee, and alumni contacts to engage them in our programming by having them advertise job opportunities to our trainees, speaking about their career paths, giving information sessions about their companies, introducing us to potential new partners, or hosting an intern. On average, approximately 50–70 in-person meetings or phone calls take place annually to build and maintain this valuable partner network.

Since the ASPIRE internship program began in 2015, over 115 trainees have participated in internships hosted by ~30 different employers in a variety of career areas that include science policy and advocacy, non-profit management, college teaching, biotechnology, healthcare data analytics, and craft beer brewing.

An additional way we have engaged partners is through *ASPIRE on the Road*, which was developed to provide trainees the opportunity to visit various companies and organizations, tour a variety of workplaces, and meet professionals in a range of roles. In May 2018, we took 11 trainees to Boston, MA, to explore Cambridge and Kendall Square. In April 2019, we took 12 trainees to San Diego, California. At each location, we visited a variety of biotechnology and pharmaceutical companies and hosted a happy hour for local alumni. We intend for this to be an annual event at different US cities.

To maintain our visibility in the community, we also serve on committees and regularly attend or speak at local conferences, especially on the topic of workforce development. For example, we are heavily engaged with the Tennessee industry advocacy organization, Life Science Tennessee.

Alumni relations and outcomes research

Alumni connections are vital to the success of the ASPIRE program, so engagement efforts with alumni continue. Twice yearly, the *Results and Discussion* newsletter — written by trainees about trainees — is published and shared with program alumni, current trainees and their parents, and our partners. This newsletter provides an opportunity for alumni to feel connected with current students and postdocs and to see their successes.

Tracking alumni career outcomes and updating contact information are essential steps in maintaining alumni engagement, as they provide information on emerging careers and ensure that program offerings remain relevant for the current employment environment. Overall training program outcomes are used to inform faculty, program administrators, institutional leaders, academic and industry partners, and alumni.

ASPIRE regularly reaches out to alumni to speak at the annual career symposia and the monthly seminar series, partner with the internship program, and meet the ASPIRE program in cities where students and administrators visit. ASPIRE hopes to continue the message that Vanderbilt trainees are valued both while they are trainees as well as afterwards, as alumni.

Faculty outreach and campus partnerships

The ASPIRE team continues to reach out to faculty via regular presentations at various types of faculty meetings. We work diligently to include faculty in as many ASPIRE activities as possible and to partner with graduate student training programs. ASPIRE leadership also meets with new faculty members and faculty who are on-boarding new postdoctoral fellows over informal lunches to share current biomedical workforce issues and professional and career development ASPIRE resources that are appropriate for the different stages of postdoctoral training.

In addition to partnerships with School of Medicine faculty, ASPIRE continues to maintain its network of campus partnerships. Initial partnerships included the Writing Studio, Department of Managerial Studies, Center for Science Outreach, CTTC, Center for Teaching, Office of Federal Relations, Clinical Microbiology and Clinical Chemistry, and the Office of Public Affairs. We also find ways to collaborate with new partners as opportunities arise. For example, we recently integrated the new Graduate School-appointed graduate student life coach in our *EQ + IQ = Career Success* module. And, with Vanderbilt's innovation center, the Wond'ry, we are finding new mechanisms for trainees to receive didactic training in entrepreneurship and commercialization.

National career development research and best practices

The ASPIRE team engages with thought leaders in biomedical research training and experts in graduate-level professional development at conferences of the AAMC GREAT group, the Graduate Career Consortium, and the National Postdoctoral Association. In the last three years (2016–19), the ASPIRE team presented over a dozen talks or posters about our module program, internships, and *ASPIRE on the Road* site visits at these conferences.

In 2016, our *Business and Management Principles for Scientists* business module, which was developed with support from a Burroughs Wellcome Fund grant and our BEST grant, was selected for an AAMC Innovations in Research Award. We subsequently published a peer-reviewed article about the module in *CBE-Life Sciences Education* [9].

Additional opportunities to share our work and collaborate with others have included the Burroughs Wellcome Fund-sponsored pre-meeting workshop of the 2017 AAMC GREAT conference, where we shared our VU ASPIRE approaches. The BEST Consortium facilitated our collaboration with other BEST institutions on the development of a taxonomy [10] to classify biomedical career outcomes, a taxonomy that was then highlighted in a 2018 pre-meeting workshop of the AAMC GREAT conference.

These conferences and professional groups provide valuable opportunities to share our program innovations and outcomes, discuss best practices, and help change the culture of biomedical research training.

Looking to the future

Balancing trainee and faculty interests with finite staff time

In its first five years, ASPIRE built a strong institutional reputation with name-brand recognition (Fig. 15.3) among Vanderbilt faculty, staff, and trainees. As the number of ASPIRE events, modules, and services has grown to its present state, it has become increasingly challenging to balance our goal of providing programming and events that can be

FIG. 15.3 The ASPIRE logo.

accessed by incoming biomedical sciences trainees with our sister goal of providing individualized trainee services. Toward this end, we continually monitor trainee event attendance, feedback, and engagement; assess the relative demand for and effectiveness of each aspect of ASPIRE; and make appropriate modifications to our schedule of events and services. Such annual strategic planning will be essential for balancing the changing demands for ASPIRE services with the finite amount of staff time available to support the program.

Interest in several aspects of our programming was very high among the trainee population when they first became available. We found that as initial demand was met, fewer trainees were interested in particular modules and events in subsequent years. This was true of entrepreneurship, technology transfer, and commercialization training; the *ASPIRE to Connect* half-day networking event; and K-12 teaching and outreach training and activities. Thus, these modules, opportunities, and events will likely be offered every other year or integrated into other programs in the future. In contrast, interest in obtaining professional headshots, oral and written communication skills building, data science training, internships at local businesses, CV/biosketch review, and *ASPIRE Café* for postdocs has remained steady or has increased. Demand for individual career counseling sessions remains very high and we anticipate allotting a significant fraction of our time to this much-requested service.

We find that the *ASPIRE Café* sessions for postdocs and presentations for graduate student training programs are time-efficient ways to provide training and information to larger groups at different stages. Thus, in the fall of 2019, we will launch *ASPIRE Bistro* — the Ph.D.-student analog to the *ASPIRE Café* for postdoctoral fellows — which will cover professional development topics tailored to the student population in a year-long series of sessions. We will also offer a new series covering the job search for both Ph.D. students and postdocs. Consolidating our existing presentations into these two new series will greatly reduce the time that staff spends leading the same sessions for multiple small groups of trainees. In addition, since existing modules and professional development sessions are now "polished" and highly rated, these will take less preparation time. With trainees experiencing introductory elements of professionalism and career training together and earlier in their training, individual career counseling can begin at a more advanced level and can thus be more time efficient. It is highly effective to provide individualized career counseling when trainees have acquired more skills and a better understanding of their values and interests, and it is also more rewarding for ASPIRE staff and faculty.

Solidifying partnerships

Developing strategic partnerships is a labor-intensive endeavor, sometimes with a long trajectory for return on investment. Nevertheless, it is a highly valuable activity that helps us grow all aspects of our programming. Although much of the motivation in identifying new partners both on campus and in our community over the past five years has been rooted in growing our internship program, we have found that many of these new connections engage with us and support our programming in additional ways, including speaking at seminars and retreats, assisting in developing new modules, posting jobs with our office, leading employer information sessions and introducing us to other potential partners.

We anticipate that the number of internship opportunities offered annually will stabilize at about 20—25 per year even as the hosts vary; many of the smaller organizations we target fluctuate in their ability or need to take on interns from year to year. Nevertheless, we will continue to go back to these hosts each year to maintain our connections with them and to inquire about internship opportunities. In many cases, due to positive past experiences, these partners are now regularly coming to us to seek interns and job candidates. Thus, we plan to let the program run organically in this manner, while adding new hosts to our repertoire as they are identified. This strategy should allow more time for staff and faculty to devote to other endeavors, such as developing new programs and providing one-on-one advising.

Finding time to design new or adopt existing programming to address the changing needs of trainees

Identifying the staff/faculty time necessary to develop and implement new programming will continue to be a challenge. To mitigate this, we will conduct careful strategic planning on an annual basis and use judicious scheduling to allow time for reflection on recent trainee feedback and preparation of new professional development sessions, modules, and large-group externships in response to the feedback we receive. Other BEST institutions have developed distinct professional development sessions that can be adopted locally; we will make use of this rich resource as we settle into a sustainable routine.

Critically analyzing and publishing outcomes

Since 2007—08, we have surveyed pre- and postdoctoral fellows at the time they complete their Vanderbilt training. These exit surveys ask trainees to report their publication, presentation, and funding successes, as well as their career interests and immediate next steps. The exit surveys also assess trainees' experiences with their program, mentor, and institutional support services, including career development. Since 2001, we have administered a "Success in Research" questionnaire to predoctoral mentors at the time their students graduate. This short, 10-item survey gauges faculty impressions of individual students. We continued our exit surveys and the "Success in Research" mentor questionnaire after ASPIRE was implemented, and together with the data gathered in the BEST Consortium entrance and exit surveys, we have rich sources of data to help us gauge the impact of ASPIRE on biomedical training.

Many elements of the ASPIRE program have been fully implemented for four years, and trainees who participated in the earlier offerings are now beginning to complete their training and move on to their next professional step. With training outcomes in hand, we will assess the impact of enhanced career and professional development programming on key parameters of research training and outcomes. Key questions we will address — either alone or in collaboration with other BEST institutions — include:

- Did participation in ASPIRE influence the time-to-degree for Ph.D. students or the time-to-completion for postdoctoral fellows?
- Did participation in ASPIRE influence trainee productivity, as measured by the number of scientific publications the trainee authored?
- How does participation in ASPIRE influence trainee career choices?
- Are graduate students who participate in ASPIRE less likely to continue on to "default" postdoctoral positions?

- Do different populations of trainees participate in career development opportunities to different degrees?
- At what point in the pre- or postdoctoral training period do trainees engage with our career exploration programs (*Ph.D. Career Connections* and the *Annual Career Symposium*)?
- Was there a shift toward participating in career exploration programs earlier in training after ASPIRE was implemented?
- Did trainee impressions of mentor/department/institutional support for career development change after ASPIRE was implemented?
- Is there a difference in mentor assessment of students who participate heavily in ASPIRE versus those who do not?
- How do internships and externships impact trainee outcomes?

We will share the results of these outcomes analyses in conference presentations and peer-reviewed publications.

Marketing successes and student satisfaction

The *Results and Discussion* newsletter is an excellent outlet to market our program's successes and student satisfaction with career and professional development opportunities, as well as a venue to celebrate the research accomplishments and activities of our trainees. Published twice a year since 2015, the full-color, 12-page newsletter is available online [11] as well as in print format. The online version is emailed to alumni as well as campus partners and administrators. Through vivid science illustrations, trainee pictures of life outside the lab, and faculty and alumni spotlights, the newsletter displays the institutional support of career programming as well as satisfied students and postdocs.

Building a grateful participant network

The BRET Office of Career Development serves as the main contact for biomedical Ph.D. and postdoctoral trainees once they leave Vanderbilt. Not only does the office ask alumni to volunteer their time at career exploration or professional development events, but it facilitates connections to current trainees. Continual dialogue with alumni also helps the office understand career outcomes better so as to inform career programs and guidance for current trainees. Because of the way the BRET Office of Career Development provides services, we hope to produce alumni who want to engage and give back because they are grateful for having received support and access to opportunities themselves, and they recognize the value it added to their training.

An exceptionally successful program we piloted in 2018 is *ASPIRE on the Road*. As described above, this initiative takes trainees to visit pharma and biotech hubs (such as Boston), as well as other venues, to highlight different types of careers. For these trips, we draw on alumni contacts to spearhead site visits at their organizations. While on the road, we invite our local alumni base to attend a networking happy hour event where they can meet the participating trainees. These events are an exciting and fun way to engage alumni who enjoy meeting trainees, reconnecting with other alumni in their city, and learning the many ways they can add value to our program by supporting current trainees.

Conclusions

The ASPIRE program leveraged an excellent foundation of career development programming and significant institutional commitment to develop a more comprehensive plan to fully prepare VUSM biomedical graduate students and postdoctoral fellows for a variety of possible careers. ASPIRE's goals have been to educate trainees about the career options that are available to them and fill any gaps in training to allow them to transition efficiently to the next step in their chosen career. Initial feedback from trainees served by ASPIRE suggests that the program has succeeded in its core mission. The sustained institutional commitment the program enjoys following the end of the five-year BEST grant suggests that the institution also recognizes the benefits of ASPIRE for trainee development. Its continued value for trainees will require maintaining a team of knowledgeable professionals dedicated to trainee success, a strong partnership with an informed faculty, an eye on the ever-changing landscape of career opportunities for biomedical scientists, and an ability to flex for the future.

Acknowledgments

We thank Angela Zito for administrative support. The ASPIRE program was developed with funding from NIGMS DP7OD018423 to Roger Chalkley, Kathleen L. Gould, and Kimberly A. Petrie.

References

[1] https://medschool.vanderbilt.edu/career-development/tag/beyond-the-lab-series/.
[2] https://medschool.vanderbilt.edu/career-development/aspire-to-connect/.
[3] https://medschool.vanderbilt.edu/career-development/annual-career-symposium/.
[4] https://medschool.vanderbilt.edu/career-development/phd-career-connections/.
[5] https://medschool.vanderbilt.edu/career-development/.
[6] https://twitter.com/vubretphdjobs?lang=en.
[7] https://twitter.com/vubretaspire?lang=en.
[8] Watts SW, Chatterjee D, Rojewski JW, Reiss CS, Baas R, Gould KL, et al. Faculty perceptions and knowledge of career development of trainees in biomedical science: what do we (think we) know? PLoS One 2019;14(1):e0210189.
[9] Petrie KA, Carnahan RH, Brown AM, Gould KL. Providing experiential business and management training for biomedical research trainees. CBE life sciences education Fall 2017;16(3). https://doi.org/10.1187/cbe.17-05-0074. pii:ar51.
[10] Mathur A, Brandt P, Chalkley R, Daniel L, Labosky P, Stayart CA, Meyers F. Evolution of a functional taxonomy of career pathways for biomedical trainees. Journal of Clinical and Translational Science 2018;2(2):63–5.
[11] https://medschool.vanderbilt.edu/career-development/results-and-discussion-newsletter/.

16

VT-BEST: Shaping biomedical professional development programming across colleges and campuses

Audra Van Wart[a, b, d], *Michael J. Friedlander*[a, b, c]

[a]Fralin Biomedical Research Institute, Roanoke, VA, United States; [b]Virginia Tech Carilion School of Medicine, Roanoke, VA, United States; [c]Department of Biological Sciences, Virginia Tech, Blacksburg, VA, United States; [d]Division of Biology and Medicine, Brown University, Providence, RI, United States

Background and organization of biomedical sciences at Virginia Tech

Building strong programming in professional development is critically important for providing doctoral and postdoctoral trainees with a solid foundation for entering a multitude of careers in the biomedical workforce. However, professional development programs, including those delivered by the Broadening Experiences in Scientific Training (BEST) awardee institutions, are not one-size-fits-all solutions. Each institution has its own culture, academic and administrative structure, and geographical advantages and limitations, all of which influence the best approach for design, implementation, participation, and sustainability. Virginia Tech has a combination of features that provide both strengths and challenges for delivering our BEST programming. At Virginia Tech, a Virginia land grant institution [1], doctoral training in the biomedical sciences is distributed across departments, colleges, and research institutes primarily located on two campuses, and degrees are administered under a single Graduate School. In this chapter, we give an overview of Virginia Tech's structure for biomedical training, the factors that influenced our BEST program design, and some general insights gained over the course of the BEST experiment.

Virginia Tech's Blacksburg and Roanoke campuses, located 40 miles apart, are connected via an intercampus shuttle that allows students and faculty to engage in seminars,

BEST: Implementing Career Development Activities for Biomedical Research Trainees
https://doi.org/10.1016/B978-0-12-820759-8.00016-4

coursework, dissertation research, and activities across the campuses. The main campus in Blacksburg houses the majority of undergraduate and graduate programming across disciplines, and has a central Graduate School that issues all Master of Science and Doctor of Philosophy degrees (conferring about 1,500 and 500 per year, respectively) and graduate certificates. Biomedical research training occurs across colleges, particularly within the College of Science, College of Engineering, College of Veterinary Medicine, College of Agriculture and Life Science, and the Biocomplexity Institute.

The Roanoke Health Science and Technology Campus houses the Virginia Tech Carilion School of Medicine, which matriculated its first students in 2010 and was recently integrated as Virginia Tech's ninth college, as well as the Fralin Biomedical Research Institute and Carilion Clinic, Virginia Tech's clinical partner. Virginia Tech's largest interdisciplinary graduate program in biomedical sciences, the Translational Biology, Medicine, and Health (TBMH) program, also operates out of the Roanoke campus, admitting approximately 20 doctoral students per year across 6 focus areas: neuroscience; cancer; immunity and infectious disease; metabolic and cardiovascular science; health implementation; and development, aging, and repair. In addition to Ph.D., M.S., and M.D.–Ph.D. students, a large percentage of Virginia Tech's postdoctoral scholars are also located in Roanoke. The rapid expansion of training in the biomedical sciences brought both a need and an opportunity to develop novel training mechanisms that could be integrated into the graduate and postdoctoral experience to better prepare them for the changing landscape of biomedical research, the diversity of research and research-related careers in the biomedical workforce, and the broad range of transferrable skills that are crucial for success across careers [2,3].

As an institution of higher learning, career and professional development have been important priorities at the university, with numerous resources dedicated to this purpose, including career advising, skills workshops, connections to internships, and certificate and degree programs (e.g., the Future Professoriate Certificate) that go beyond the major training and are provided through the Graduate School's Transformative Graduate Education Initiative [4]. Despite this, by the time we launched Virginia Tech's BEST Program, VT-BEST, it was clear that there was still a substantial lack of resources specifically geared toward graduate and postdoctoral trainees in biomedical and health sciences and that there wasn't yet a strong push to expand traditional research training to better prepare trainees for the variety of career paths they might encounter within and beyond academia. Further, there was a growing need for better data collection about graduate students and postdocs at Virginia Tech to gauge whether or not they had an interest in expanded training opportunities, such as workshops and internships. We also needed to improve the tracking of career outcomes for trainees in the biomedical sciences.

To build support for such initiatives through VT-BEST and promote culture change, we took an initial approach of informing trainees, faculty, and administrators about the NIH Biomedical Workforce Working Group Report published by the NIH [2] and the origins of the NIH Director's Biomedical Research Workforce Innovation Award: Broadening Experiences in Scientific Training (BEST) [7,8]. In particular, we highlighted the findings that the majority of US-trained biomedical Ph.D.s were employed in careers other than tenure-track faculty positions and discussed the proposed strategy of VT-BEST for addressing the report's recommendations.

Engaging faculty and program planning

At the inception of VT-BEST, we held two faculty forums where we presented the findings of the NIH report [2] and where attendees had an open discussion about the role of our graduate and postdoctoral training programs in preparing trainees for the changing biomedical workforce needs and the expansion of the breadth of career opportunities within and beyond academia. We also issued an anonymous survey to them about their perceived capacity to advise trainees on non-academic careers (Fig. 16.1) and their support and perceptions of the time their trainees spend on professional development activities (Fig. 16.2). Following these discussions and the administration of the faculty survey, it became clear that the best approach for achieving broad trainee participation and faculty support for career development activities would be to provide activities that occupied smaller intervals of time and that exposed trainees to experiences and networks beyond those available to them through their mentors and curricula alone.

As the VT-BEST program relied on faculty to assist with designing and delivering new content (often in collaboration with non-academic professionals), connecting it to their professional networks, hosting guest speakers, allowing their trainees to participate, advocating for

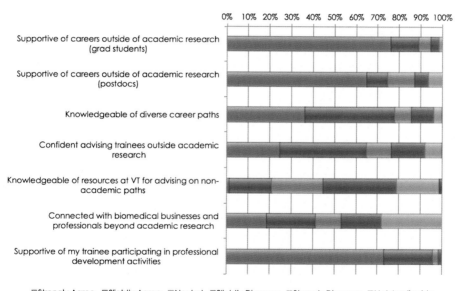

FIG. 16.1 Survey of faculty support for non-academic careers (N = 77). In February 2014, we held two open faculty forums for about 80 faculty members, one for 60 faculty and stakeholders from departments with interest in the biomedical and health sciences at the main Virginia Tech campus in Blacksburg, and one for 20 faculty located at the Fralin Biomedical Research Institute at the Roanoke medical campus. The attending faculty completed a brief anonymous survey to assess baseline faculty opinions and attitudes. Overall, faculty were widely supportive of their trainees pursuing careers beyond academic research, however, their capacity and confidence for advising them along these trajectories was limited, highlighting the need for activities that provide mentorship and exposure to different career opportunities.

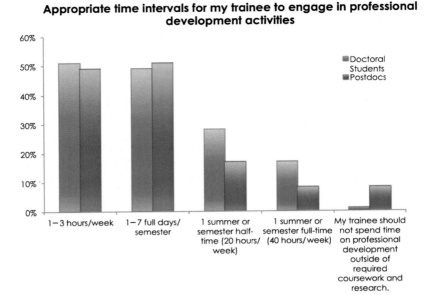

FIG. 16.2   Faculty support for the time trainees spend on professional development (N = 77). Although faculty were overwhelmingly supportive of their trainee participating in professional development activities, opinions were mixed regarding the appropriate amount of time to spend on such activities. Support for longer intervals was much lower, particularly for postdoctoral trainees.

it, and being part of a culture change at Virginia Tech, faculty buy-in was particularly important for the experiment to succeed. Importantly, the program also required that the many participating departments and graduate programs regularly and thoroughly collect data and assist in encouraging both BEST and non-BEST trainees to complete periodic surveys. These data were required for the assessment of the BEST Consortium program coordinated by the NIH and for internal formative and summative assessments of VT-BEST. While the university collects some basic information in a standard fashion (such as admissions and graduation statistics), VT-BEST efforts required frequent and more detailed data collection on current and incoming individual trainees. Although it was at times challenging to gather information that the university does not collect in a centralized manner, doing so gave us the opportunity to connect regularly with representatives from programs across the university.

VT-BEST has a director who oversees its functioning, a coordinator who organizes activities, and an evaluator who develops the data collection strategy. The program also made use of an advisory board that includes both internal and external members from graduate education and industry. Each year, the program engages a core set of faculty members who facilitate instruction, advise on program elements, support and encourage trainee participation, and tap into their professional networks to identify appropriate individuals from outside Virginia Tech and the Virginia Tech alumni network who can help deliver content. These VT-BEST PIs and faculty reach out to partners in the region who can advise and participate, including representatives from Virginia Tech's Corporate Research Center (home to over 185 research, technology, and support companies), the Roanoke Blacksburg Technology Council, Virginia Bio (Virginia's premier, statewide non-profit trade association for the life sciences industry, which has over 250 member organizations), and other industry partners in the region.

The program was also influenced by the early assessment of a cohort of graduate students who attended pilots of our Individual Development Planning (IDP) workshops. The pilot attendees represented 10 different departments, from Biomedical Sciences and Pathobiology to Human Nutrition, Foods, and Exercise. Through the assessment, we learned about the broad career interests of the students at our institution and found that they were most interested in learning about research careers across multiple settings (government, academia, industry). It also became clear from reflective discussions with the students that the IDP workshops were very helpful, but would be more valued if they were embedded in a course that included information on careers, building of professional skills, skills-assessment activities, and direction toward in-depth resources that could be pursued selectively and strategically.

Program structure, implementation, and feedback

One important aim for VT-BEST is to allow the broadest participation possible and to not be limited to one specific cohort. Thus, we opted for an *à la carte* model that permitted all trainees the opportunity to participate (or apply for participation) in the full range of VT-BEST activities. Although this approach does not provide structured career and professional guidance for each VT-BEST participant on a longitudinal basis, it does allow them more opportunities to attend one-off events, which in the end led to a greater inclusion of postdocs and graduate trainees from both campuses.

Curricular components

VT-BEST centers all programming around the IDP. In the abovementioned series of IDP-planning pilot workshops and related surveys, it became clear that while trainees had the greatest interest in learning about research career opportunities across settings, there was also significant interest in exploring a broad range of career paths; thus, we aimed to feature a wide range of careers in our VT-BEST programming. The IDP pilots formed the foundation of a Professional Development and Ethics course, which was rooted in the TBMH Graduate Program but was open to all trainees interested in taking the course for credit, auditing the full course, or sitting in on specific sessions. The presented content was a combination of career-oriented and skill-building topics, such as preparing the IDP, grantsmanship, managing lab finances, communicating science, media training, and career paths for Ph.D.s. Trainees were required to complete several deliverables by the end of the course: performing one informational interview, attending one non-BEST professional development activity, completing the IDP, and reviewing the IDP with a mentor, among other tasks.

Since one of the VT-BEST PIs is also a founding co-director of the interdepartmental TBMH Graduate Program, TBMH frequently serves as a testing ground for integrating VT-BEST program elements more fully into the graduate training experience. For example, although each department sets its own procedures for the Graduate School-required annual assessment of its students, the TBMH Graduate Program has modified the assessment to include a section dedicated to annual professional development goals that go beyond research training and publication number. While the IDP remains a personal working document that the students shape with guidance from their mentorship teams, which are made up of any faculty or external professionals that the students view to be their professional mentors, the professional development component of the TBMH assessment formalizes key annual goals that the mentor and mentee agree to prioritize each year. As TBMH mentors often train students from numerous other graduate programs, we hope that they continue to apply this practice with all of their trainees.

Interestingly, VT-BEST has also provided an opportunity for faculty to test new ways of integrating material directly into the basic science coursework. For instance, TBMH runs an eight-credit course in the fall that covers the foundations of translational biology, medicine, and health and describes successful and unsuccessful exemplars of translation from different programmatic focus areas (e.g., neuroscience, cancer, infectious disease). VT-BEST supported the efforts of program faculty from the Biomedical Engineering and Mechanics Department to incorporate a commercialization module, which culminates in a "shark tank" pitch competition, directly into the course. For the shark tank pitch event, students present ideas for the development, investment, and commercialization of products in the health sciences space for which intellectual property has already been developed through the Virginia Tech Intellectual Properties (VTIP) Office.

Module participants include students from the Pamplin College of Business and the Virginia Tech-Wake Forest School of Biomedical Engineering. Six bioscience professionals provide a total of 12 hours of instruction and mentoring on topics such as intellectual property, regulatory considerations, and startups to prepare the students for the project. The students, working in teams of four to six individuals who represent a diversity of backgrounds (e.g., economics, engineering, biology, chemistry, psychology, nursing, and computation), have several weeks to complete the projects. The shark tank presentations are judged by a panel of judges that includes investors, intellectual property attorneys, biotech startup entrepreneurs, business managers, and scientists. The trainee teams address key considerations such as the breadth of the technology, market need, investment strategies, regulatory issues, and return on investment.

This module has evolved each year with great success. It has served as a launching pad for trainees to integrate a broader worldview into their research projects and has exposed them to a variety of paths they can pursue as scientists. Indeed, some of the groups have gone on to form their own startups and win larger statewide pitch competitions.

Extra-curricular components

Outside of the classroom, VT-BEST has experimented with other mechanisms for exposing trainees to different career experiences. One component of VT-BEST programming that has been quite successful used a format we dubbed "mini-internships." These experiences were designed to be job-simulation workshops and were inspired by the rigorous and extensive job interviews that are common in industries such as academic publishing and business consulting. Under the guidance of Virginia Tech faculty, mini-internships are led by external Ph.D. professionals (in some cases, by M.D.s, J.D.s, or M.B.A.s) from various career paths who prepare original exercises that are designed to take trainees through scenarios or job tasks that represent their job responsibilities. The goal is for trainees to get a sense of the skills and knowledge required for the featured job, identify areas where they need to improve, and reflect on whether the career type fits in with their abilities, strengths, and values. At the end of each mini-internship, trainees have time for reflection, questions, and networking, such that inspired trainees can get a sense of the next steps for learning more about that career and for developing the skills they need should they choose to pursue it.

Most mini-internship sessions have been hosted at Virginia Tech, although some have involved traveling to given job sites. Often, mini-internship instructors are invited to stay in town for an extra day to give a more traditional career talk, part of another VT-BEST programming component. We found that enthusiasm for mini-internships, which are far less expensive and less time intensive than more traditional externships, shadowing, and internship experiences, is extremely high.

Although arranging internships was not an original aim of the VT-BEST program, internship opportunities often arose through the development of relationships between the Roanoke Health Science and Technology Campus and industry partners, and VT-BEST became known as a contact point for companies looking to advertise internships. In these cases, VT-BEST program staff helped to match appropriate trainees to such opportunities and sometimes even helped to formalize or structure these arrangements. We found that the majority of trainees who complete internships train with academic mentors who are generally supportive of industry careers and who are engaged in translational research themselves. The majority of these trainees are also near the end of their training and are looking for industry experience before applying for jobs.

Beyond these core elements, VT-BEST faculty, staff, and trainees integrated a variety of other activities into each year's programming. Examples include career talks from Ph.D.-level scientists that represent a variety of job sectors (publishing, government, industry, academia, communications, policy/non-profit, etc.), a trainee-driven career blog, and the formation of a career resource library featuring books reviewed by Virginia Tech trainees. We have also delivered numerous skills workshops in partnership with other Virginia Tech units, such as the Center for Communicating Science, the Graduate School, the Career and Professional Development Office, and various departments and graduate student organizations.

We have also set up short, multi-day trips to biotech and small pharma companies in our region. This allows our trainees to compare and experience the diversity of job settings, structures, cultures, and business models represented in the biotech sector, and to understand the opportunities that are available in the region. Finally, we sponsor a travel award program that trainees can apply for to help offset the costs of attending professional development activities, such as internships, job shadowing, skills workshops, and networking opportunities targeted to particular employment sectors. Trainees receiving individual funds for such opportunities are required to either give a presentation upon their return or write a blog post about their experiences.

Program sustainability through integration and partnership

A new office at Virginia Tech will provide continued funding not only for BEST programming but for further experiments and innovation in the area of biomedical training.

Fortunately, the VT-BEST program development has occurred in parallel with the growth of Virginia Tech's Health Science Campus (Roanoke) and with the expansion of graduate training programs. Virginia Tech established the Office of the Vice President (VP) for Health Sciences and Technology as well as various health science initiatives that span both campuses during the NIH BEST grant period. The office, which was founded in 2016 to manage the growth and strategic planning of the Roanoke campus, is funded through the Office of the Provost and will support continued VT-BEST programming through financial and personnel resources. The VP for Health Sciences and Technology oversees the research, clinical, education, and outreach visions of the office, engages industry partners and other stakeholders, and works to attract new bioscience businesses to the region. The Assistant VP for Health Science Education, also housed in this office, focuses on biomedical education and training efforts. Currently, VT-BEST leadership fills both of these roles, which means that they are well positioned to continue the oversight of VT-BEST activities under the domain of the new office. They are also poised to facilitate the development of new training activities that are consistent with the BEST mission.

It is additionally worth noting that a new research and education building will open on the Health Science and Technology Campus in 2020. The new building will provide additional space for educational programming, including VT-BEST activities, such as scale-up classrooms, experiential learning studios, collaborative learning labs for commercialization projects and hackathons, and data visualization spaces.

A second method for sustaining VT-BEST post-NIH grant is incorporating various programming components into existing structures through collaborations with different graduate programs, units, and partners at Virginia Tech. For example, Virginia Tech's largest interdisciplinary graduate program, TBMH, is now responsible for the coordination and funding of some VT-BEST courses, workshops, and seminars, which have been integrated into the graduate program administration and budget. These offerings include the Professional Development and Ethics course, IDP training, and associated expenses for the speakers in courses and some of the career talks. The Office of the Provost has approved funding for these activities, and several TBMH faculty already contribute their time and efforts toward them.

One of the programs adopted by TBMH is the VT-BEST commercialization and translation module, which has participating students from within the TBMH structure, as well as from biomedical engineering and business. Due to the success of the module, we are planning to develop a second full course in partnership with other units at Virginia Tech that will have a similar structure but that will provide a more intensive experience for a more limited number of students. In the future, the course may include medical students and undergraduates, as they are also important contributors to the biomedical workforce.

Throughout the grant award period, VT-BEST partnered with other Virginia Tech offices and external entities, which will now help maintain the continued delivery of programming and other opportunities. For example, VT-BEST has worked with the Career and Professional Development Office to deliver activities such as networking training sessions before professional/scientific events. The office has now hired additional staff that is focused on graduate student-related topics who can also assist in forging collaborative efforts, such as Ph.D.-focused career fairs and events, between colleges and campuses. The Graduate School, which established a university-wide initiative called Transformative Graduate Education [7], also supports the development and launch of new, non-degree classes that target graduate training and professional development needs across disciplines. Virginia Tech's Center for Communicating Science has also worked closely with VT-BEST to expand its offerings across both campuses, including through the delivery of special workshops (on visualizing data and social media for scientists, for example) and integration of smaller lessons (such as using improve to build communication skills) into the Professional Development and Ethics class. The university has since provided further investment in the center to broaden the range of coursework and opportunities available across the two campuses.

Beyond Virginia Tech, our recent partnership with RAMP [8], a local accelerator, has provided training in commercialization and startups at all levels. Last summer, graduate students and postdocs participated for the first time in parallel commercialization training through workshops held at RAMP and through shadowing opportunities held as the new companies progressed through the RAMP program. This new annual summer program is one example of the type of training activities that are growing at Virginia Tech whose coordination is facilitated by existing infrastructure.

Lastly, a key component of sustainability is the trainees themselves. Over the past five years, there has been a growing level of student and postdoc interest in coordinating various activities, ranging from inviting guest speakers to planning an annual career and professional development day. VT-BEST staff have worked closely with the Roanoke Graduate Student Association, Virginia Tech Carilion Student Outreach Program, and individual students and postdocs to either collaborate or advise on the planning of trainee-driven professional development activities. Moving forward, this relationship will help reduce the staff time required for event planning — leading to a greater number of offered events — and will promote continued trainee participation in these peer-led professional development efforts.

Challenges, lessons, and advice

While the university structure and the rapid growth of biomedical and health sciences research have been an advantage at Virginia Tech, there are also factors that have presented challenges. The distribution of our trainees across two campuses challenges us to offer activities that can reach them across both sites. In response to this need, some events are duplicated at both sites (such as IDP workshops) while others are web-streamed or transmitted through interactive videoconferencing. Ultimately, the most effective and cost-efficient mechanism has been to encourage trainees to make the trek to the campus where a particular activity is held. We found that the best way to encourage this was to plan our events around regularly scheduled shuttle times or to coordinate dedicated transportation for each activity. Having a pre-registration or application selection process, a pre-event assignment, and regular communications prior to the activity promoted better participation. Another effective approach was to have partners and ambassadors on both campuses participate in events at the other campus and encourage trainees to make the trek as well.

A second challenge has been targeting our activities and data collection efforts to biomedical trainees, which can be difficult considering the fact that all graduate training is integrated into a single graduate school that houses diverse programs that range from history to biology. Although our Graduate School is overwhelmingly supportive of students and their career outcomes, the centralized delivery of professional development and the nature of the data that need to be collected must be broadly appropriate rather than being discipline specific. This creates a challenge not only for delivering specialized programming and career guidance, but also for aspects such as outcomes tracking, where there is not a single, agreed-upon career taxonomy for collecting outcomes information centrally across all degree programs. Accessing and extracting data solely about biomedical trainees is also a challenge, as much of the centralized data collection efforts aggregate data from colleges and departments that have diverse research interests and disciplines. Individual-level data are collected to various extents by each department, but this is not consistent across the units that house biomedical trainees. Thus, we used central resources to announce programming and pull aggregate data whenever possible, but relied on departments participating in VT-BEST to commit to the data collection approach outlined in the program proposal.

Virginia Tech has a relatively small number of biomedical sciences postdocs (approximately 70−100 in the departments and institutes participating in VT-BEST); our postdocs are categorized differently from student trainees and enjoy similar benefits and services as faculty and staff (including access to courses taught through the Virginia Tech Career and Professional Development Office). This population, however, can be more challenging to engage and connect with, considering that the university does not have an active postdoctoral association or a central office dedicated to postdocs. VT-BEST worked to identify and include as many biomedical sciences postdocs from across the university as possible, and made efforts to include them in a thoughtful way, such as by having them nominate and host outside professionals to present at events or career talks. In recent years, our Roanoke Graduate Student Association has been particularly encouraging of postdoc participation in professional development activities such as the annual Professional Development Symposium.

Another consideration for institutions interested in BEST-like programming is that, like Virginia Tech, not every institution is located in an industry hub. However, our area, like many others, is not without opportunity. There is still a close-knit and lively community of bioscience industry professionals that represent companies from across the Roanoke and Blacksburg region. Virginia Tech also has an award-winning Corporate Research Center that has just expanded to accommodate growth and that is developing a growing, prestigious research park for high-technology companies seeking to advance the research, educational and technology transfer missions of the university. Additionally, the Blacksburg and Roanoke campuses are not far from Washington, DC, which facilitates trainee participation and involvement in non-profit, science policy, government organizations (such as the NIH and the FDA), or larger biotech and pharma companies; these externship or informational interview experiences can include single- and multi-day trips. Even without residing in an industry hub, Virginia Tech has access to an active alumni network that is useful for identifying professionals who are enthusiastic about leading VT-BEST activities.

We have learned a few things about maximizing participation through the timing and scope of activities. Although trainees regularly express a strong interest in attending a variety of career and professional development activities, the reality is that securing good attendance requires instilling a notion of relevance and a balance between frequency and scope of topics. For us, the greatest success comes when students are involved in the planning process and get commitments from their peers to participate. Activities offered early in graduate training are best received when they emphasize actions students can immediately take to begin developing strong CVs, skills, and networks. Wherever appropriate, we have used students or postdocs who are near the end of their training to provide a reflection on the actions they took (or wish they had taken) in the early phases of their training; these reflections have typically been well received. We have also received particularly positive feedback from our experiential learning activities that involve more hands-on elements, such as our mini-internships, as trainees feel that they provide significant insight into what a given job is like. Finally, we advise putting mechanisms into place to encourage regular reflection and goal setting and to direct trainees toward relevant resources. These could be institutionalized by incorporating them into required programming or into annual academic assessments, as we have done with our TBMH trainees.

Conclusions

For Virginia Tech, the NIH BEST Award provided an important opportunity to test new training mechanisms, exchange best practices across a consortium of institutions, increase communication and collaboration across our own institution, and secure critical buy-in for reshaping our training programs for graduate students and postdocs. Our approach for integrating the delivery and support of VT-BEST content into Virginia Tech's growing biomedical and health sciences infrastructure leaves us well prepared to carry forward existing programming, shape the growth of future programs designed to enhance the training experience, and respond to the evolving needs of the biomedical workforce.

Acknowledgments

We would like to acknowledge our hard-working VT-BEST program staff and all of the faculty and guest speakers who have contributed to the program content. We are also grateful to our advisory board members and our colleagues across the BEST Consortium, who have provided helpful discussions and collaborations. Lastly, we would like to acknowledge our VT-BEST graduate students and postdocs for their participation in the program and for their engagement in shaping the activities and the training culture at Virginia Tech. VT-BEST was supported by NIH grant DP7OD018428.

References

[1] https://www.nap.edu/read/4980/chapter/2.
[2] National Institutes of Health. Biomedical Research Workforce Working Group Report. Bethesda, MD, USA: National Institutes of Health; 2012.
[3] Sinche M, Layton RL, Brandt PD, O'Connell AB, Hall JD, Freeman AM, Harrell JR, Cook JG, Brennwald PJ. An evidence-based evaluation of transferrable skills and job satisfaction for science PhDs. PLoS One 2017;12(9):e0185023.
[4] Available from: https://graduateschool.vt.edu/transformative-graduate-education-experience/tge-initiative.html.
[5] National Institutes of Health. NIH Director's biomedical workforce innovation award: broadening experiences in scientific training (BEST) funding opportunity announcement RFA-RM-12-022. Bethesda, MD: National Institutes of Health; 2013.
[6] Meyers FJ, Mathur A, Fuhrmann CN, O'Brien TC, Wefes I, Labosky PA, Duncan DAS, August A, Feig A, Gould KL, Friedlander MJ, Schaffer CB, Van Wart A, Chalkley R. The origin and implementation of the Broadening Experiences in Scientific Training programs: an NIH common fund initiative. The FASEB Journal 2015;30(2):507−14.
[7] https://graduateschool.vt.edu/transformative-graduate-education-experience/tge-initiative.html.
[8] https://ramprb.tech/.

Across disciplines: Multi-phase career preparation for doctoral students

Christine S. Chow[a], Ambika Mathur[b, c],
Judith A. Moldenhauer[d]

[a]Department of Chemistry, Wayne State University, Detroit, MI, United States; [b]Department of Pediatrics, Wayne State University (previous), Detroit, MI, United States; [c]University of Texas, San Antonio, TX, United States; [d]Department of Art and Art History, Wayne State University, Detroit, MI, United States

Institutional context and overarching goals of the Wayne State University Broadening Experiences in Scientific Training program

Wayne State University (WSU) is a major comprehensive research institution located in Midtown Detroit, MI. The university has a large medical school and a strong urban mission that aims to meet the needs of the local population. The WSU Graduate School oversees approximately 1,500 doctoral students and 300 postdoctoral associates, about 400 of whom are in biomedical fields in 15 programs spread out across 10 (out of a total of 13) schools and colleges.

As with the majority of research-intensive institutions, WSU's doctoral programs traditionally focused on preparing students for postdoctoral research and careers in academia despite the fact that only ~25% of Ph.D. graduates go into tenure-track faculty positions [1]. Thus, the majority of graduates were left to navigate other career paths themselves. Our recognition of this gap motivated us to create a program to transform biomedical training so that students could learn about and prepare for multiple career pathways. We strived to have the highest level of institutional commitment and, most importantly, strong support from faculty and alumni.

As depicted in the logic model (Fig. 17.1), the WSU Broadening Experiences in Scientific Training (BEST) program was developed with the **assumptions** that: (1) while biomedical doctoral students receive excellent training in disciplinary research they do not receive adequate training in exploration and preparation for careers beyond academia; (2) although

Logic Model – WSU BEST

ASSUMPTIONS

- Most biomedical doctoral students become proficient in research but do not have career training that can open doors to other career opportunities.

- Individual doctoral programs do not have appropriate knowledge and/or resources for providing information about careers beyond academia.

- Institutional changes made for biomedical disciplines can positively impact students in other degree programs (halo effect).

RESOURCES

Wayne State University (WSU) has 400 doctoral students in biomedical graduate programs.

WSU Graduate School has a structure to provide graduate seminars for career development (e.g., GPPD).

WSU has faculty and professional partners to develop curricula for workshops/seminars and experiential learning opportunities in five different career paths: business, communication, government, law, and teaching.

STUDENT ACTIVITIES

- *Outreach* to incoming doctoral students

- *Individual Development Plan (IDP)* tailored to career development

- *One-hour career exploration seminars* in the five career paths

- *Full-day career preparation workshops* in the five career paths

- *Mentored experiential learning opportunities* with an employer in one of the career paths

- *Additional workshops and seminars (GPPD)* about transferable skills and professional development opportunities that transcend career paths

OUTPUTS

- *100% of doctoral students* completed an IDP.

- *14% of biomedical doctoral students*, on average, participated in the annual orientation session.

- *19% of biomedical doctoral students* participated in one or more of the five 1–2 hour exploratory seminar modules, each focused on a different career path.

- *7% of biomedical doctoral students* participated in one or more of the five full-day interactive career preparation workshops.

- *44 individual doctoral students* participated in experiential learning.

LEARNING OUTCOMES

1 Doctoral students demonstrate *increased awareness* of career sectors (academic and non-academic). (*Method: post-session student surveys*)	2 Doctoral students demonstrate *increased knowledge* of skill sets necessary for success in these careers. (*Method: post-session student surveys*)	3 Doctoral students have *increased opportunities* to explore careers beyond academia. (*Method: post-session student surveys*)	4 Doctoral students outside of the biomedical field have *opportunity, guidance, and support* to pursue diverse careers in addition to academia (halo effect). (*Method: post-session student surveys*)	5 Graduate students explore careers indepth using *experiential learning*. (*Measure: number of non-academic experiential learning placements*)

SHORT-TERM GOAL

Doctoral (biomedical and non-biomedical) students will have a *greater interest and intent to pursue diverse careers in addition to academia*.
(*Method: post-session student surveys*)

LONG-TERM GOAL

More doctoral students *are placed in careers* in addtion to academia.
(*Method: tracking of post-graduation career placement*)

FIG. 17.1   Logic model for the WSU BEST program.

individual graduate programs provide excellent discipline-specific training to their doctoral students, they generally lack the skills, knowledge, and resources to provide their doctoral students with training and preparation for careers beyond academia; and (3) programs created to provide such knowledge to biomedical doctoral students could be extended to doctoral students in other disciplines, thus producing a "halo effect" that would benefit all students.

The WSU Graduate School recognized that it would be burdensome for each individual graduate program to plan its own activities, and thus used the BEST program as a tool to deliver unified programming. These **activities** include the mandatory completion of an Individual Development Plan (IDP) for all doctoral students, a three-phase career exploration, and Graduate and Postdoctoral Professional Development (GPPD) seminars and workshops. The expected **learning outcomes** for doctoral students in the program are that: (1) they will demonstrate *increased awareness* of career sectors; (2) they will demonstrate *increased knowledge* of transferable skill sets required to succeed in all careers; (3) they will have *increased opportunities* to explore careers beyond academia; (4) they will have *opportunity, guidance, and support* to pursue diverse careers in addition to academia even if they are outside the biomedical field; and (5) they will explore careers in depth using *experiential learning*.

Although the BEST program was designed for doctoral students in biomedical graduate programs, we appreciated the need to include all Ph.D. students and postdocs in our programming. Therefore, a unique feature of the WSU BEST program is that all of our events are open to all doctoral students and postdocs. As indicated in Fig. 17.1, the *short-term* goal of the program was to provide new opportunities and guidance for doctoral students to **explore** a wide variety of career paths while learning about the **skill sets** needed for various job sectors and exploring specific career paths through **experiential learning**. The *long-term* goal of the program was to institutionalize these practices so that students could enter the workforce in their desired areas **with intent**, rather than taking haphazard paths towards those careers. Given WSU's urban mission, another *long-term* goal was to partner and continue **building relationships** with local industries and businesses (e.g., non-profit organizations, government agencies, law firms, primarily undergraduate institutions, etc.) to not only bring strength to our program, but to also create a talent **pipeline** into those professions.

Rationale and need for the program

We began the program at WSU by first assessing the career outcomes of previous doctoral graduates. In 2013, we only had data for certain programs, such as those with training grants. Our initial analysis showed that, between 2008 and 2012, fewer than 25% of our graduates pursued academic positions—numbers that are consistent with the national trends [1−3]. It was clear we needed new programming for current graduate students and long-term data tracking of all our graduate alumni. However, we lacked a mechanism to assist the faculty and administration in addressing the inadequate training models of career development. Our program was designed to address these programming gaps by exposing students to a wide variety of career paths, giving them professional development exposure, and training them in skills that would supplement their discipline-specific research experiences.

Institutional support

We made the decision to house the WSU BEST program in the Graduate School for a number of reasons. First, having a centralized unit to run the programming is more efficient than having individual units each providing their own career development options. Second, the Graduate School provides us with staffing and funding support, particularly for experiential learning, which occurs in *Phase III* of the program (more details below). Third, the Graduate School is able to reach all doctoral students, postdocs, and faculty because of its central role in the university; its relationship with schools, colleges, departments, and centers; and its access to WSU communication outlets. Fourth, having resources centralized through the Graduate School allows faculty and individual units or programs to maintain their focus on discipline-specific training while directing their students to use the career resources offered through the BEST program and the Graduate School's professional development activities. Our approach was intentional, with the hope of having strong faculty buy-in and transforming attitudes across the university to accept the importance of career development for our doctoral student population.

Program participants

The WSU Graduate School admits approximately 150 students annually into its biomedical doctoral programs. These students are from 15 "target" departments and programs (Anatomy, Biochemistry and Molecular Biology, Biological Sciences, Biomedical Engineering, Cancer Biology, Chemistry, Communication Sciences and Disorders, Immunology and Microbiology, Molecular Genetics and Genomics, Nutrition and Food Science, Pathology, Pharmacy, Pharmacology, Physiology, and Translational Neuroscience). However, since WSU is a mid-sized institution, we were able to create programming that was attractive to and that included all departments ("non-target") on campus (e.g., History, Physics, Political Science, Psychology, Sociology). Although we initially designed the programming for incoming doctoral students (those in their first year), we found that the highest number of participants are post-candidacy doctoral students who are closer to graduation (those in their third year and beyond).

Since 2013, the WSU BEST program has involved more than 1,000 unique participants (for reporting purposes, we define each individual as a "unique participant" regardless of how many activities they attend), including approximately 800 doctoral students, 50 postdocs, 100 faculty (as workshop facilitators, panelists, etc.), 100 panelists from various career sectors, 15 internal advisory committee members, 15 external committee members, 40 industry partners, several graduate student assistants, 2 external evaluators, and a program manager. We also offer our programming to students from neighboring institutions; therefore, several events have also included doctoral students from the University of Toledo, Michigan State University, and the University of Michigan.

Program structure

As shown in Fig. 17.2, we developed three main career blocks for BEST that focus on career planning, career preparation, and institutional transformation and broader impacts.

WSU BEST Program Career Blocks

Block A   **Career Planning**

Orientation
Discussion of careers with research mentors using IDP and SciPhD

Block B   **Career Preparation** (with bi-directional faculty engagement)

Three-phase career preparation (experiential learning of job skills)
    Phase I – Introduction to Careers
    Phase II – Interactive Career Preparation Workshops
    Phase III – In-depth Experiential Learning

Competency-based professional development (transferable soft skills)
    Communication
    Leadership and Professionalism
    Teamwork and Collaboration

Block C   **Institutional Transformations and Broader Impacts**

Acceptance of multiple career pathways
Evaluations
Scalability
Dissemination of outcomes and processes

FIG. 17.2   Overview of WSU BEST's model of career blocks.

Block A comprises an orientation for first-year graduate students as well as any students or postdocs who have not previously participated in the program. The Graduate School provides opportunities for career planning and development through the IDP. The IDP became a requirement of all WSU doctoral students in 2014, as it encourages discussions between students and their faculty advisors about the students' educational goals. WSU holds several workshops annually for students and faculty that focus on the implementation of IDPs.

Block B comprises two parts that were developed in collaboration with the Graduate School. The first part is a three-phase career exploration and experiential learning series (BEST *Phases I, II,* and *III*). The second part is the GPPD seminar series. The BEST *Phases I, II,* and *III* focus on specific skills of the individual career tracks, whereas GPPD presentations provide insight on skills and techniques that students can broadly apply to their research and professional activities. For part two of Block B, the Graduate School created a set of competencies linked to transferable skills that are applicable across disciplines, along with a digital micro-credentialing system for students to track and demonstrate acquisition of these skills [4].

Block C involves the evaluation of program outcomes and the dissemination of that information both within and outside the university. Such widespread dissemination increases buy-in, guides program improvements, and institutionalizes successful practices, all of which benefits our trainees. The success of Block C is evident; as of 2019 (after the completion of the

BEST funding), the WSU Graduate School still organized and ran GPPDs and BEST events and individual departments still held workshops that follow the BEST format (referred to as "miniBEST" events).

The three phases of career preparation

A pyramid structure, illustrated in Fig. 17.3, is used for career preparation (Block B). We introduce all incoming graduate students to the BEST program during the first week of new-student orientation. Students then have the opportunity to participate in 60- to 90-minute sessions (*Phase I*) that focus on five career tracks (described in the next section); there are no limitations on the number of sessions that students can attend. In the first year of the program, these sessions were videotaped and are available online to all students at any time. *Phase I* is intended for all doctoral students, but is also open to master's students, post-docs, faculty, staff, and undergraduate students. Each *Phase I* session includes a panel discussion that highlights career opportunities and provides knowledge about the expectations and skill sets needed for a professional working in that career track. The sessions include panels composed of WSU alumni, program partners (e.g., from industry, non-profits, government, etc.), and a faculty moderator.

Following *Phase I* participation (either attendance at a session or viewing of the corresponding video), students can attend *Phase II*, in which they engage in day-long interactive career workshops focused on building skills and gaining knowledge about careers in specific tracks. As with *Phase I*, there are no limitations on the number of *Phase II* sessions that trainees can attend. These seven-to eight-hour workshops include a variety of activities, such as interactive team projects and student presentations to the entire group of students and workshop facilitators. The faculty moderators for these workshops hold conference calls with the presenters (e.g., alumni, industry partners, etc.) in advance to develop the workshop content, learning outcomes, schedule, and assessments for the activities, which can include informal feedback for the trainees and evaluations of the events.

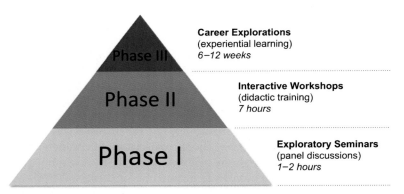

FIG. 17.3 Overview of the WSU BEST three-phase career exploration program. The level of knowledge, programming focus, and time required increases from *Phase I* (bottom) to *III* (top). The highest participation occurs in the short exploratory *Phase I* and is reduced in *Phase II*, which involves a structured, full-day approach to specific careers. The lowest participation rate occurs in the experiential *Phase III*, which requires on average a 160-hour time commitment with an identified external partner.

Each *Phase II* workshop includes three or four different activities that provide the students with some background about day-to-day job duties in a particular career sector. Students also see how they can apply some of their current discipline-specific skills to other areas and learn what skill sets they need to develop to succeed in that career track. Some workshop activities have included, for example, writing a press release on a current topic for lay audiences on social media or for a journal highlight (Science Communication track), preparing a public policy vote recommendation for a government official (Government track), developing a business plan (Industry/Business track), or using student-centered teaching methods to present to introductory biochemistry undergraduate classes (Teaching at Primarily Undergraduate Institutions [PUIs] track). Once the workshop presenter introduces the background on each session's career topic and skill, the participants work in small teams to complete the exercise and to present to the entire group. These experiences allow students to develop and share ideas with experts in the field and participants from other disciplines on campus [5].

Phase III provides students an in-depth, several weeks-long work experience with a company or organization whose focus matches one of the career tracks from *Phases I* and *II*. The duration of the experience is typically 160 hours and occurs during the summer. Applications to participate in *Phase III* of the BEST program require a brief essay, a transcript, an up-to-date IDP, and approval by the trainee's faculty research mentor and their department's Director of Graduate Studies. Upon completion of *Phase III*, the student participants are required to submit a report about their experience. Although the WSU BEST program was designed primarily for doctoral students, postdocs are also able to participate in *Phase III*.

Program activities

To shift the training paradigm of WSU doctoral programs, WSU BEST provides students with the resources to explore multiple career paths. Prior to 2013, the training of WSU doctoral students primarily involved preparation to enter postdoctoral positions as a stepping stone to then go on to academia or industry. The BEST program is structured to provide richer opportunities for students to develop skills through a combination of didactic programs, experiential learning, workshops, and in-depth career explorations. As illustrated in Fig. 17.4, WSU BEST was designed strategically to prepare students for careers in sectors that we identified through an alumni census [4], and offers career tracks in Business/Industry, Communication, Government, Law/Regulatory Affairs, and Teaching at PUIs. In addition to these, we have offered other tracks such as Community Engagement in the first year and Research Administration in the fifth year of BEST. GPPDs provide students with additional career planning and preparation.

Program leadership and guidance

WSU BEST put together a steering committee that met weekly to design and implement programming. The committee was composed of the Director of the Office of Teaching and Learning, two external evaluators, a graduate student representative, and eight faculty and administrators from the School of Medicine; the College of Liberal Arts and Sciences; the

FIG. 17.4 WSU BEST program career preparation and outcomes. The upper half (red) highlights the activities related to career exploration and preparation; the lower half (black) shows career sectors supported by our program. The "Academia" sector refers to research-intensive positions, in contrast to the teaching focus favored at PUIs.

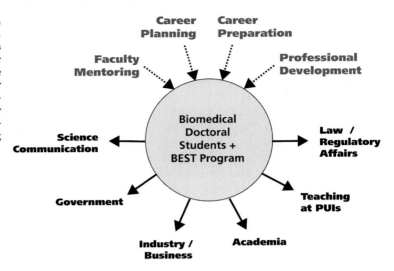

College of Fine, Performing, and Communication Arts; the College of Education; the School of Social Work; and the College of Pharmacy. This inclusive approach was important for changing the institutional culture and gaining broad acceptance of career planning across the entire university. The BEST steering committee assisted with program development, assessment, and student selection for experiential learning.

WSU hired a BEST Program Manager who played a key role in data collection and reporting, as well as in assisting with day-to-day activities such as handling budget issues, coordination of events, marketing of the program, and identifying and communicating with internship partners.

In the first two years of the WSU BEST program, we held several workshops and seminars for faculty to expose them to what we do, which was important for the faculty to develop career programming within their own departments and for the steering committee to gather ideas for the development and improvement of *Phases I, II*, and *III* and the GPPDs.

The internal advisory committee, composed of faculty and administrators from across the university who were not directly involved with BEST, met annually and provided oversight throughout the program development. They helped with institutional commitment, increasing faculty engagement and buy-in, and finding industrial partners.

The external advisory committee was composed of alumni, *Phase III* partners, and industry representatives and played a role in giving the program a better understanding of the skill sets needed by students and assisted with program planning. The BEST program also partnered with the Office of Teaching and Learning to develop learning outcomes for each track in the three phases.

Program outcomes

WSU's Institutional Review Board on the Use of Human Subjects approved all research conducted in this program (IRB#094013B3E). SPEC Associates, a third-party non-profit research and evaluation organization based in Detroit, performed all the evaluations. SPEC Associates obtained demographic information on BEST participants and nonparticipants (the comparison group) from WSU and Graduate School records. They obtained students' departmental affiliations from registration and survey records. All data are reported in aggregate form or with identifiable information removed.

At the beginning of the funding period, WSU carried out a census of approximately 3,000 doctoral alumni who graduated between 1999 and 2014. The results, similar to those reported by others [2], indicated that biomedical doctoral alumni work in a variety of job sectors, including industry/business (31%), academia (tenured/tenure-track, 29%), teaching at PUIs (13%), government and law (5%), science communication (<1%), ongoing training such as postdoctoral positions (16%), and more [4,6].

In the fall of 2013, the BEST program began *Phase III* research experiences for a small group of students. In 2014, all three *Phases (I–III)* were held for five different career tracks in addition to a year-long series of GPPDs. Overall, participation during the first three years of BEST programming (2014–17) included 44% of all doctoral students in the biomedical departments, and only 1–15% of all doctoral students in non-biomedical departments. By 2018, more than 150 unique students from biomedical departments (and over 50 from non-biomedical departments) participated per year, such that, over the course of their entire graduate career, about half of the biomedical students engaged in career development, each attending a range of 1–17 events.

We analyzed the BEST program outcomes to determine whether there were different participation rates throughout our career programming and whether there were differences in the career outcomes of women and students from underrepresented backgrounds [5], but did not observe any statistically significant differences with regards to gender, race, or citizenship. We did, however, find correlations in which students who participated in BEST activities tended to have higher incoming GRE scores and GPAs. In contrast, there was no correlation between participation in BEST and time-to-degree, even when the students participated in five or more events [5].

We evaluated the outcomes of all WSU BEST programming elements (*Phases I–III* and GPPDs) based on student awareness of career options (including academia), access to opportunities, and interest in pursuing different careers, as well as guidance and support available to them for pursuing diverse career tracks. Our external evaluators (SPEC Associates) carried out formative and summative evaluations using a Retrospective Pretest (RPT) methodology. This analysis was important to determine changes from before and after BEST participation, as well as to assess the effectiveness of the individual events [5]. The comments from students on variables such as career topics, timing, and length of events were particularly helpful in altering programming, as was the case with the abovementioned Research Administration and Community Engagement tracks, also discussed below.

Lessons learned

An important feature of the WSU BEST program is the "halo effect" it has had, evidenced by the fact that doctoral students from a wide range of "non-target" departments (particularly Physics and Astronomy, Communication, and History) have participated in our events.

This halo effect has positively impacted our sustainability, since faculty and administrators have seen the campus-wide student response to this program and are now committed to the long-term institutional support of the programming. The inclusivity of our program has added further benefits, such as the fostering of cross-disciplinary interactions between our students; most students do not get the opportunity to interact with students outside of their discipline-specific training and may not see the benefit of having diverse teams to solve problems or to accomplish tasks in real-world settings. Thus, the students from the arts, education, humanities, physical sciences, and social sciences contribute significantly to these biomedical training programs. We believe that centralization of the career programs in the Graduate School helps to facilitate this broad participation.

We did find (as did other institutions) that participation in individual events can be quite variable and can result in students experiencing "career development fatigue." For example, *Phase I* and *II* participation rates for the Law/Regulatory Affairs and Government tracks tend to be much lower than for Industry/Business, Communication, and Teaching, which is not surprising given the career paths of our graduates. Therefore, it may be better to hold some *Phase I* sessions less frequently (e.g., once a year instead of once a semester) so that the participation rates are high enough to justify the time and cost of bringing in panelists and facilitators. For instance, we have held two *Phase II* workshops (each covering a separate subject) in a single day, such that the morning and afternoon sessions shared a combined lunch and panel discussion. This structure helps minimize students' time off from research.

Student comments on *Phase I* surveys and panel discussions are used to tweak the programming, which also helps with participation rates. For example, students reflected their desire to learn more about the topic of research administration, so we introduced this as a track in 2017. In contrast, the Community Engagement workshop had such low attendance in year one of WSU BEST that we eliminated it from our list of tracks. However, we found that students appreciated having community engagement topics built into the other workshops and incorporated them thusly. A Community Engagement GPPD (a one-hour session) was also created as an alternative and has had much higher attendance than the full-day workshop.

We found that student participation varied, but with some correlation to the duration and intensity of programming. Although some students participate in numerous BEST and GPPD workshops and seminars, the majority of students benefit from involvement in just a few events, particularly high-impact events that cover many topics over the span of a few days, such as the SciPhD workshop [7]. Other students prefer to participate in career development programming offered through their home departments, so WSU BEST was involved in the establishment of "miniBEST" programs. The cost of running a smaller event is typically $2,000 to $4,000, depending on the number of participants and length of the event (e.g., hour-long events vs. day-long events) with a high return on investment. Most of the WSU "miniBEST" events involve alumni from the specific academic program, and some departments have received additional financial support through the development/philanthropy offices of their schools and colleges.

In summary, the WSU BEST program has several key features that students and faculty both appreciate and that other institutions can replicate. We developed centralized programming that provides students the ability to explore multiple career paths and to gain knowledge about the skill sets needed for success in various career sectors. We have found that doctoral student participation is broad—coming from all disciplines on campus—with equal attendance regardless of gender, race, or citizenship status. Furthermore, participation in career development activities does not impact the time-to-degree-completion, which was an initial concern of our faculty. The career-focused activities of the BEST program and our approach to integrating those activities with the students' required academic training have resulted in yearly increases in buy-in from the faculty and administration. Although centralization helps with sustainability, the challenge moving forward, as at other institutions, will be in maintaining financial support of the program as well as continuing to track data and reporting student outcomes.

Acknowledgments

We thank the WSU trainees, faculty, staff, alumni, steering committee, and all of the internal and external partners who helped us develop and deliver content for the WSU BEST program. We also thank the internal and external advisory committee members for their valuable input and guidance. We are grateful to the National Institutes of Health (DP7OD01827) and the WSU Graduate School for funding.

References

[1] NIH, National Institutes of Health Biomedical research workforce working group report, Bethesda, MD; 2012. https://acd.od.nih.gov/documents/reports/Biomedical_research_wgreport.pdf.

[2] Meyers FJ, Mathur A, Fuhrmann CN, O'Brien TC, Wefes I, Labosky PA, Duncan DS, August A, Feig A, Gould KL, Friedlander M, B Schaffer C, Van Wart A, Chalkley R. The origin and implementation of the Broadening Experiences in Scientific Training programs: an NIH common fund initiative. The FASEB Journal 2016;30:507—14.

[3] Hitchcock P, Mathur A, Bennett J, Cameron P, Chow C, Clifford P, Duvoisin R, Feig A, Finneran K, Klotz DM, McGee R, O'Riordan M, Pfund C, Pickett C, Schwartz N, Street NE, Watkins E, Wiest J, Engelke D. The future of graduate and postdoctoral training in the biosciences. Elife 2017;6.

[4] Mathur A, Cano A, Kohl M, Muthunayake NS, Vaidyanathan P, Wood ME, Ziyad M. Visualization of gender, race, citizenship and academic performance in association with career outcomes of 15-year biomedical doctoral alumni at a public research university. PLoS One 2018;13:e0197473.

[5] Mathur A, Chow CS, Feig AL, Kenaga H, Moldenhauer JA, Muthunayake NS, Ouellett ML, Pence LE, Straub V. Exposure to multiple career pathways by biomedical doctoral students at a public research university. PLoS One 2018;13:e0199720.

[6] Feig AL, Robinson L, Yan S, Byrd M, Mathur A. Using longitudinal data on career outcomes to promote improvements and diversity in graduate education. Change 2016;48:42—9.

[7] https://SciPhD.com.

Index

Note: Page numbers followed by "t" indicate tables, "f" indicate figures, and "b" indicate boxes.

Printed in the United States
By Bookmasters